Make a Raspberry Pi–Controlled Robot

Wolfram Donat

MAKER MEDIA
SEBASTOPOL, CA

Make a Raspberry Pi–Controlled Robot

by Wolfram Donat

Printed in the United States of America.

Published by Maker Media, Inc., 1005 Gravenstein Highway North, Sebastopol, CA 95472.

Maker Media books may be purchased for educational, business, or sales promotional use. Online editions are also available for most titles (*http://safaribooksonline.com*). For more information, contact O'Reilly Media's corporate/institutional sales department: 800-998-9938 or *corporate@oreilly.com*.

Editor: Patrick Di Justo
Production Editor: Melanie Yarbrough
Copyeditor: Sharon Wilkey
Proofreader: Kim Cofer
Indexer: Angela Howard
Cover Designer: Riley Wilkinson
Interior Designer: Nellie McKesson
Illustrator: Rebecca Demarest

November 2014: First Edition

Revision History for the First Edition:

2014-11-10: First release

See *http://oreilly.com/catalog/errata.csp?isbn=9781457186035* for release details.

ISBN: 978-1-457-18603-5

[LSI]

Table of Contents

Preface

So you want to build a robot.

Like many others before you, you saw the introduction of the Raspberry Pi minicomputer as a milestone in not just portable computing technology, but mobile robotic technology. After all, here was a device the size of a credit card, with a processor equivalent to a Pentium III. Here was a device about the same size as an Arduino board, but capable of HD 1080p graphics. Here was a 700MHz CPU, with a set of 26 GPIO pins that could connect it to the outside world. Here, in a nutshell, was a robotic brain.

Unfortunately, it probably didn't take you very long to figure out that *calling* a computer a robotic brain and *making* it a robotic brain are two very, very different things. Sure, you can plug the Pi into your desktop monitor, add a keyboard and a mouse, and start to program. But unless you have a specific goal in mind and a clear path to get there, it's easy to get lost in the mechanics of writing the program, and adding sensors and motors and switches and cameras and the countless other things that make up a robot. So even though the Raspberry Pi made robotics cheaper and smaller, it didn't necessarily immediately make it easier (see Figure P-1).

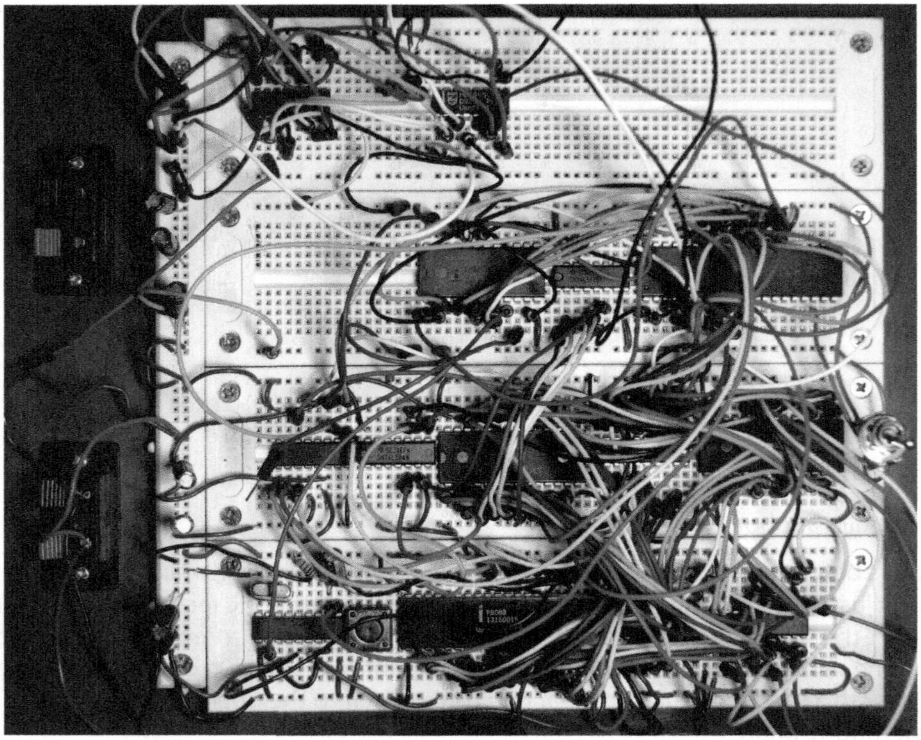

Figure P-1. *Not as easy as it looks*

We're fortunate that this didn't stop people from pressing forward; a Google search for "raspberry pi robot" brings up over two million results, not counting the YouTube videos and all of the specialized subsearches, like "raspberry pi robot arm" and "raspberry pi robot servo" and "raspberry pi robot butler." Robotics is just like any other discipline: there will always be backyard tinkerers, hobbyists, and off-duty professionals, as well as hardware and software hackers who take off-the-shelf parts and stretch them to—and past—their limits.

Yes, the swarm of synchronized flying drones at MIT is awesome, but that project has several million dollars in funding behind it. Meanwhile, your next-door neighbor has succeeded in creating a wheeled robot that can play fetch with his dog, chase the kids, and play *Minecraft*, all for a budget of under $500—something I find quite a bit more impressive.

Conventions Used in This Book

The following typographical conventions are used in this book:

Italic
: Indicates new terms, URLs, email addresses, filenames, and file extensions.

`Constant width`
: Used for program listings, as well as within paragraphs to refer to program elements such as variable or function names, databases, data types, environment variables, statements, and keywords.

`Constant width bold`
: Shows commands or other text that should be typed literally by the user.

`Constant width italic`
: Shows text that should be replaced with user-supplied values or by values determined by context.

> *This element signifies a tip, suggestion, or general note.*

> *This element indicates a warning or caution.*

Using Code Examples

This book is here to help you get your job done. In general, you may use the code in this book in your programs and documentation. You do not need to contact us for permission unless you're reproducing a significant portion of the code. For example, writing a program that uses several chunks of code from this book does not require permission. Selling or distributing a CD-ROM of examples from Make: books does require permission. Answering a question by citing this book and quoting example code does not require permission. Incorporating a significant amount of example code from this book into your product's documentation does require permission.

We appreciate, but do not require, attribution. An attribution usually includes the title, author, publisher, and ISBN. For example: "*Make a Raspberry Pi–Controlled Robot* by Wolfram Donat (Maker Media). Copyright 2015 Wolfram Donat, 978-1-4571-8603-5."

If you feel your use of code examples falls outside fair use or the permission given here, feel free to contact us at *bookpermissions@makermedia.com*.

Safari® Books Online

Safari Books Online is an on-demand digital library that delivers expert content in both book and video form from the world's leading authors in technology and business.

Technology professionals, software developers, web designers, and business and creative professionals use Safari Books Online as their primary resource for research, problem solving, learning, and certification training.

Safari Books Online offers a range of product mixes and pricing programs for organizations, government agencies, and individuals. Subscribers have access to thousands of books, training videos, and prepublication manuscripts in one fully searchable database from publishers like Maker Media, O'Reilly Media, Prentice Hall Professional, Addison-Wesley Professional, Microsoft Press, Sams, Que, Peachpit Press, Focal Press, Cisco Press, John Wiley & Sons, Syngress, Morgan Kaufmann, IBM Redbooks, Packt, Adobe Press, FT Press, Apress, Manning, New Riders, McGraw-Hill, Jones & Bartlett, Course Technology, and dozens more. For more information about Safari Books Online, please visit us online.

How to Contact Us

Please address comments and questions concerning this book to the publisher:

Make:
1005 Gravenstein Highway North
Sebastopol, CA 95472
800-998-9938 (in the United States or Canada)
707-829-0515 (international or local)
707-829-0104 (fax)

Make: unites, inspires, informs, and entertains a growing community of resourceful people who undertake amazing projects in their backyards, basements, and garages. Make: celebrates your right to tweak, hack, and bend any technology to your will. The Make: audience continues to be a growing culture and community that believes in bettering ourselves, our environment, our educational system—our entire world. This is much more than an audience, it's a worldwide movement that Make: is leading—we call it the Maker Movement.

For more information about Make:, visit us online:

Make: magazine: *http://makezine.com/magazine/*
Maker Faire: *http://makerfaire.com*
Makezine.com: *http://makezine.com*
Maker Shed: *http://makershed.com/*

We have a web page for this book, where we list errata, examples, and any additional information. You can access this page at *http://bit.ly/make_a_raspberry_pi_controlled_robot*.

To comment or ask technical questions about this book, send email to: *bookquestions@oreilly.com*

Acknowledgments

It's true—writing a book is a solitary endeavor, but it definitely can't be done alone. There are several people I'd like to thank, without whom neither the Rover nor this book about it would exist.

First and foremost, thank you to Becky and Reed, for supporting me and putting up with my parts, tools, and half-built, weirdo projects scattered around the house.

Dexter and Jörgen—you can both stop talking now.

And finally, a big thanks to all of the crew at Make: and O'Reilly—Brian, Melanie, Frank, Sharon, Gretchen, Dale, and especially Patrick. You guys are awesome, and all of your support was much appreciated!

Introduction 1

The task of building a robot is unlike any other in computer science. It's a strange amalgamation of computer, electrical, and mechanical engineering. Being able to program is great (and necessary), but if you can't get your program to interact with physical items like sensors and motors, then your robot will forever be a virtual one. If your motors can't move your rover without pulling more current than your circuits can source, the rover will be immobile until you find a solution —either different motors, a different circuit, or a lighter rover. And using a reed switch to determine when your rover runs into a wall is a great idea, until you discover that the switch you bought online can't stand up to the force of a 20-pound rover hitting a wall at 10 miles per hour. You need to learn to roll with the punches, fix what breaks, and—when possible—prevent it from breaking in the first place.

Building a robot also requires knowing your limits, related to both your knowledge and your materials. I *really* wanted to put a robotic gripper hand on this rover, and chances are I will eventually, but I'm aware that it probably won't happen without different tools and better materials than those I can find at the corner hardware store. Likewise, the ion-drive engine is going to have to wait a few years; in the meantime, electric car-seat motors will have to do. And be prepared to know and accept when one of your designs is just *wrong*, and to go back and redesign something. By following along in this book, hopefully you'll be taking advantage of my making the mistakes for you; rest assured that the rover design you see in this book is by no means the original design I had in my head, though I am pretty happy with the results.

The flip side to knowing your limits, of course, is being willing to stretch those limits when you think you can, and to be ready to think of unconventional ways to do things, especially when you're a backyard tinkerer—a Maker. PVC pipe, for example, is meant to be used for plumbing. However, it also makes excellent shock-absorbing drive axles (see Chapter 7). Yes, I'm using plumbing flex-hose to cover the guts of my robotic arm, and the rover's wheels are pulled straight off a Power Wheels vehicle. Sometimes you can experience great flashes of inspiration just by wandering the aisles of your local hardware (or toy) store. Sometimes you can solve a particularly knotty problem the same way.

I like to call this robot a *rover*, as I tried to pattern it after NASA's designs. Figure 1-1 shows the general outline of the finished rover.

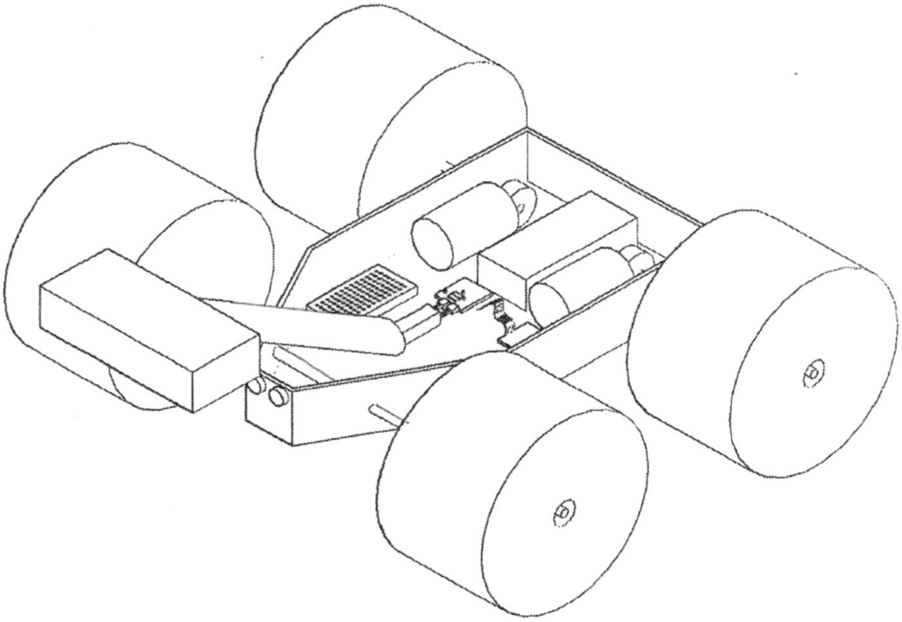

Figure 1-1. *The rover*

It's not nearly as robust as NASA's versions, of course, and you'll notice that its four (not six) wheels don't sit on their own independent shock absorbers, but the design is a proven one. And speaking of wheels: although I would very much like to program my own anthropomorphic android, such as C-3PO, it's a sad fact that the Raspberry Pi's computing power is most likely not up to the task of controlling a bipedal droid. You may think it's nothing special, but as it happens, getting a robot to not only balance on two legs, but also walk on them, is quite a challenge. The well-known ASIMO robot by Honda (Figure 1-2) required many years and many millions of dollars to finally be able to walk on its own.

To balance on two feet, a robot's internal sensors must constantly measure where the robot's center of gravity (COG) is, and then determine where the robot's feet are, and then check to see that the COG is over at least one of the robot's feet, preferably over a line between the robot's feet, or at most, very slightly offset from that line (but not too far). If the robot's COG is too far to one side, the robot's brain must send the command to flex the leg on that side to tilt the robot ever so slightly in the other direction, bringing the COG to a more stable location, without going too far in the other direction. And if the robot is carrying something, all those values need to be recomputed on the fly.

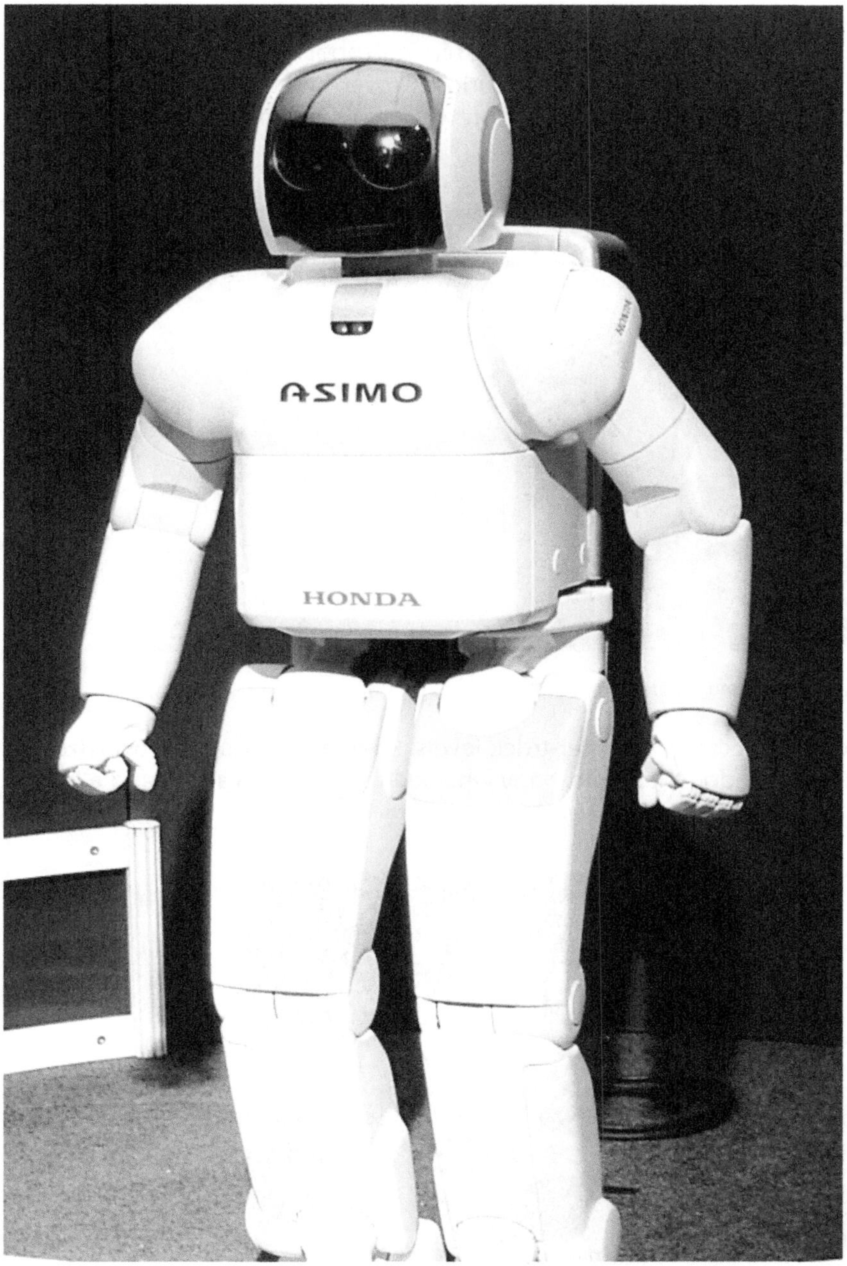

Figure 1-2. *More than the Pi can handle*

So there are several advantages to using wheels. First, not having to balance means that the Pi's computing power (and servo power) can be spared for other tasks, such as taking temperature samples or moving the robot arm. Second, depending on the type of wheels you use, a wheeled vehicle can go all sorts of places that a bipedal robot can't. And third, wheels can also be cool—I refer you to R2-D2, the Mars Curiosity rover, and the Mars Exploration rovers (Spirit and Opportunity) for examples of pretty cool wheeled robots. Figure 1-3 shows the Mars rovers.

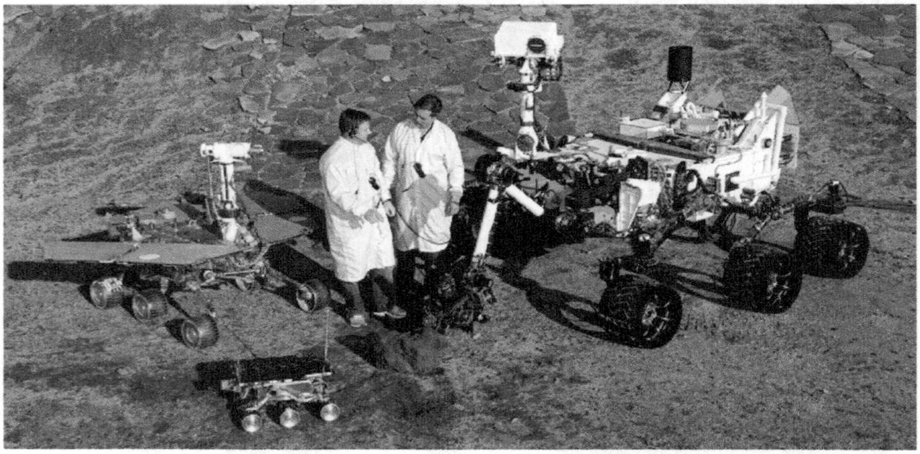

Figure 1-3. *Three bad-assed wheeled robots*

To increase the coolness factor to monster-truck levels, I decided to go with oversized wheels; it's common knowledge that almost any wheeled vehicle looks seven and a half times better with bigger tires. Figures 1-4 and 1-5 prove my point.

Figure 1-4. *Small tires: not so awesome*

Figure 1-5. *Big tires: AWESOME!*

This brings up more design challenges, however. Larger wheels tend to be heavier, and it's always—*always*—a good idea to keep your robot or rover as light as possible. A heavy robot is a power-hungry robot, and batteries and engines are heavy enough to begin with. Large wheels also have greater rolling resistance, though rolling resistance comes more into play at higher speeds and higher efficiencies than this rover is likely to experience. My solution: I used the wheels from a Power Wheels vehicle. They're large and impressive, but because they're made of plastic, they hardly weigh anything. Of course, that led to further challenges, such as mounting those wheels to a non–Power Wheels axle, but as you'll see in Chapter 7, those issues were solved as well, often with a combination of screws, nuts, bolts, and generous applications of epoxy and cold-weld.

The final design, assuming you follow these step-by-step instructions, can be seen in Figure 1-6.

Figure 1-6. *Finished rover*

As you build it, of course, you'll need to keep pace programming it to interact with the sensors you choose to use with it. I do all of my Pi work in Python, and that's what you'll see in this book. If you're not familiar with this powerful, more-than-a-scripting language, flip to Appendix B for a quick introduction. The Pi was designed to run Python, and most of the libraries and modules you need are a quick `sudo apt-get` command away, with a few exceptions that I'll walk you through as you need them. In Chapter 10, as I introduce you to some sensors, I'll also show you the Python code necessary to work with each one. In most cases, the code you write to test the sensor can be saved and used as a function (`getTemperature()`, for instance) in the final rover code. In Chapter 11, I'll give you a final working program, but if you've been following along, you'll have written all the bits and pieces already, so it won't be anything new to you.

As you go through this book, following along breathlessly as I impart this robot-building wisdom, keep one thing in mind: these build instructions are meant to be *suggestions*. If you can't find the same aluminum channel I used for the robotic arm, or if you have an idea that you feel would work better, by all means, use it! And then tell me about it! I always look forward to seeing what my fellow builders come up with.

If you're ready, let's get started by taking a look at the Raspberry Pi.

2 Intro to the Raspberry Pi

Because the robot we're building is using the Raspberry Pi for its brain, giving you a short introduction to this nifty little computer isn't a horrible idea. If you're already familiar with it, feel free to skip this chapter. Otherwise, read on for a quick tour around the board (Figure 2-1). As of this writing, the Raspberry Pi Foundation has released a new version of the board, called the model B+, with extra USB ports and more GPIO pins. If you have one of these newer boards, I'll go over its features a bit later in the chapter.

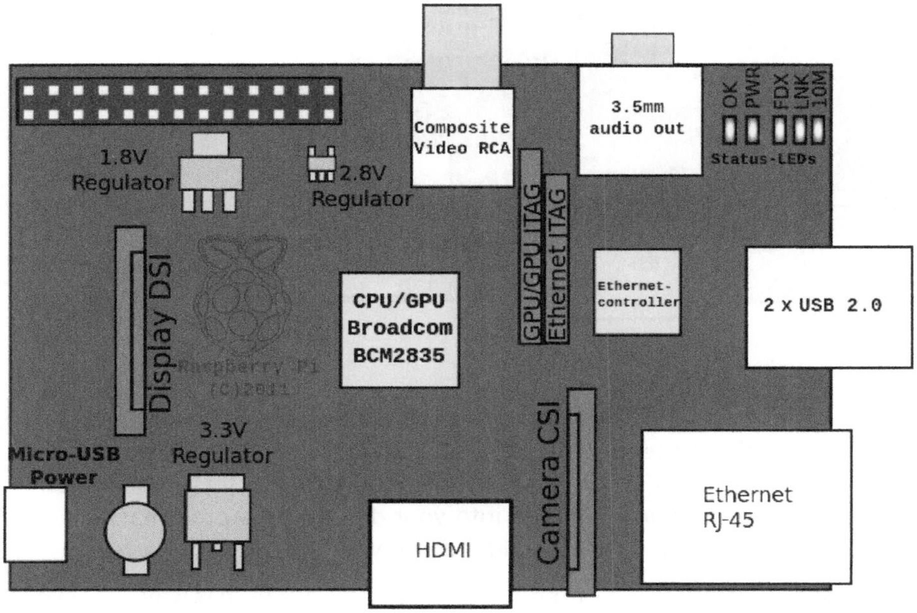

Figure 2-1. *The Raspberry Pi*

Model A and Model B

We can start the tour at the Ethernet port, as that's pretty much common ground for any computer you're familiar with. It's a standard 10/100 port—nothing special about it at all. If you're not familiar with the terminology, the "10/100" stands for 10 and 100 megabits per second (Mbps), the two worldwide standard data rates for Ethernet communication. Older computers are limited to 10Mbps, and newer computers are capable of 100Mbps, but they can usually communicate at slower speeds to maintain backward compatibility. All Ethernet networks are connected to a central hub or switch, and that connection is either wireless or via twisted-pair cables attached to an RJ-45 connector that closely resembles an old analog phone jack.

Moving counterclockwise around the board, the next thing you come across is the pair of USB ports (Figure 2-2).

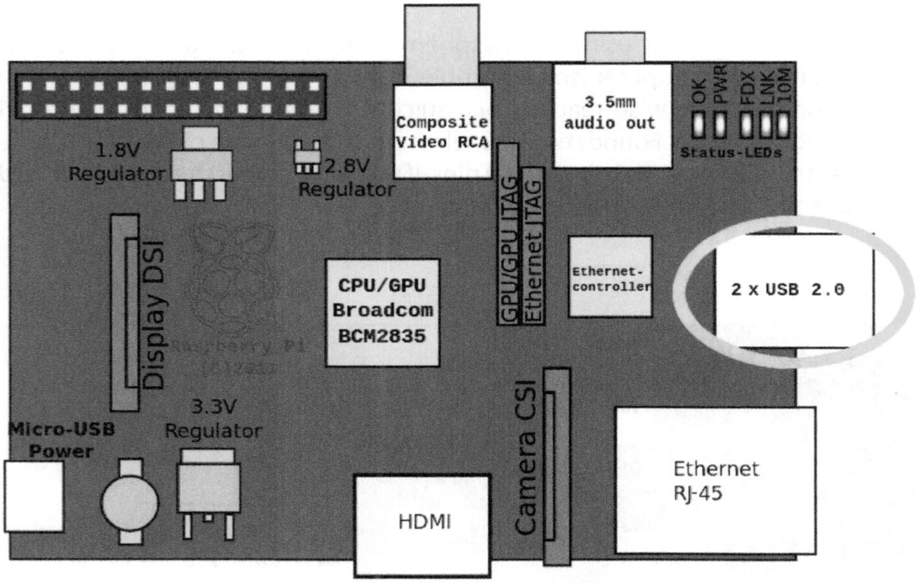

Figure 2-2. *USB ports*

Both these USB ports and the LAN (Ethernet) port are handled via the onboard LAN9512 chip. According to the datasheet, the chip is capable of 480 Mbps USB 2.0 speeds, and fully integrated 10base-T and 100-baseTX Ethernet support. What that means, for lack of a better description, is that almost any device you plug into your desktop or laptop machine—printer, external hard drive, USB fan—can be plugged into your Pi.

These ports let you plug in a keyboard or mouse, and control your Pi that way. You can even plug in a USB hub to connect more devices to your Pi. This is a common configuration if you use a WiFi USB dongle; you can plug the dongle into one of the Pi's USB ports, and plug your USB hub (Figure 2-3) into the other. (Don't try to plug a WiFi USB dongle into a USB hub—you'll most likely get strange behavior. Some devices need to be connected directly to the Pi.)

If you use a USB hub, get one that's externally powered. The Pi doesn't provide a lot of current, and relying on it to power both a hub and the connected devices can lead to even more strange behavior from your Pi and the devices. On the simplest level, the devices might not even get enough power to work. On the other hand, the hub I use is not externally powered, and it works fine, so your results may vary.

Figure 2-3. *Belkin F5U407 mini USB hub*

The next step on the path around the board is the row of five status lights (Figure 2-4). In order, from the center of the board outward, they're labeled OK (or ACT, if you have the Pi version 2.0), PWR, FDX, LNK, and 10M (or 100), and are green, red, green, green, and orange, respectively. The FDX and LNK lights may be orange, again depending on your board version.

Figure 2-4. *Status lights*

These lights can be helpful for troubleshooting your Pi. Because the Pi doesn't have a BIOS like most computers, nothing gets printed to the screen if there's a boot failure, leaving you to interpret the lights. The green ACT light flickers when there is SD card activity, such as writing or reading to memory. It should always be a bright green when it's lit; a dull green glow means that no boot code has ever been executed. The red PWR light means that the board has 3.3V and is powered properly. The FDX and LNK lights are related to connectivity: FDX means there is a full-duplex Ethernet connection, and LNK means that there is activity on that connection. Finally, the 100 light means that there is a 100 Mbit Ethernet connection.

Moving beyond the status lights, we come to the audio jack. It takes a standard 3.5mm headphone plug. Next to it is the composite video RCA jack (Figure 2-5), where you can connect to an external video device—such as a pair of video goggles like the MyVu set. The Pi doesn't support RGB video, unfortunately; connecting it to a monitor will require a monitor with an HDMI port.

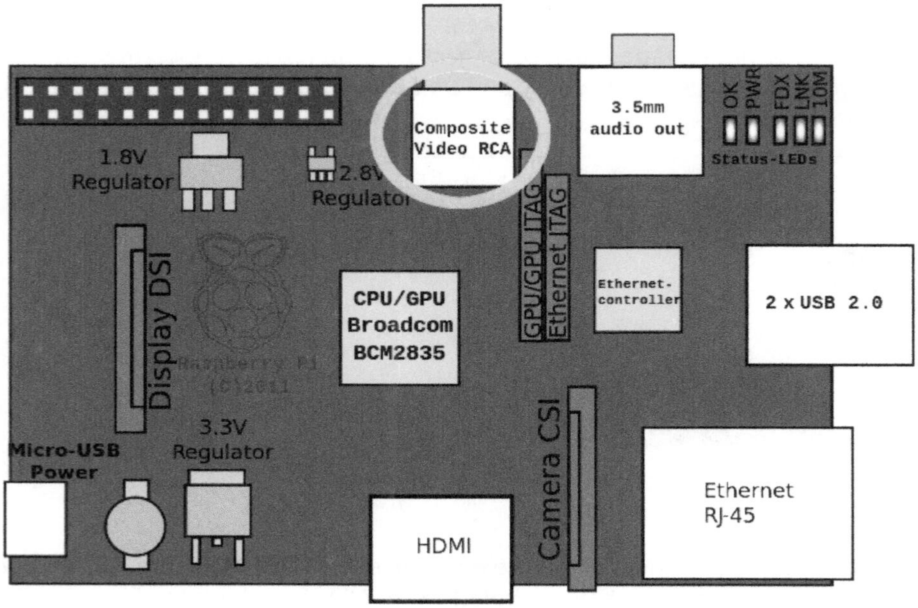

Figure 2-5. *Video RCA jack*

The next stop on our tour is arguably the coolest thing about the Pi. The two rows of pins sticking straight up are the general-purpose input/output (GPIO) pins (Figure 2-6). They enable the Pi to interact with the physical world—getting input from sensors and controlling outputs like motors, servos, and lights. You may remember when laptops and desktops had serial and parallel ports, which could be used, with a little effort, as interfaces to the computer hardware. They've pretty much been replaced by USB ports. With the Pi, we have a computer that gives us access to the hardware again. Using the GPIO pins, you can immediately control at least eight servos—enough for a quadruped robot, for example.

Using the Python *RPi.GPIO* library, which is included in later versions of the Pi's Raspbian operating system, we can turn specific pins into INPUTs or OUTPUTs. If you've used the Arduino integrated development environment (IDE) at all, you'll recognize the concept. With the Arduino, to set up a pin as OUTPUT and send voltage to it, you use the following:

```
pinMode(11, OUTPUT)
digitalWrite(11, HIGH)
```

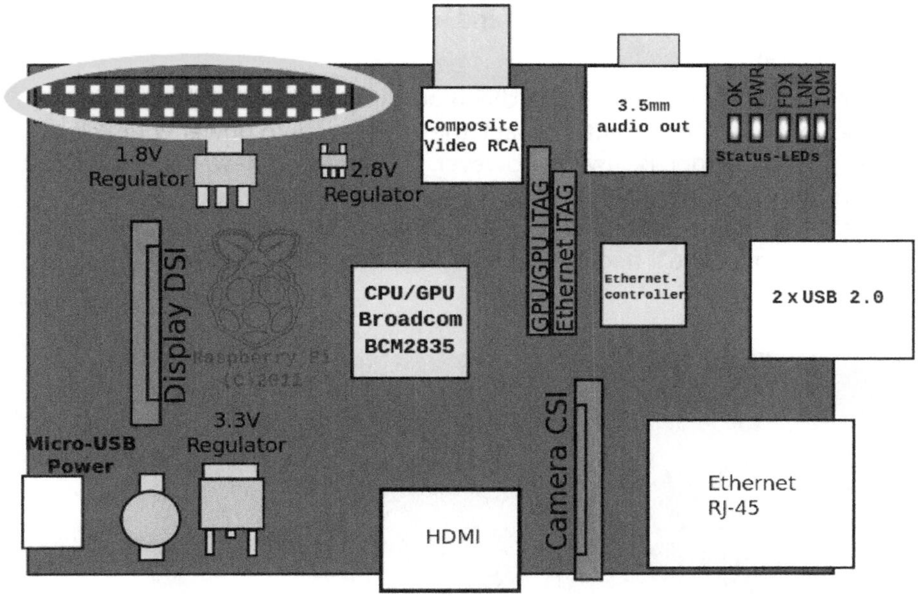

Figure 2-6. *GPIO pins*

With the Raspberry Pi, you use this:

```
import RPi.GPIO as GPIO
GPIO.setmode (GPIO.BCM)
GPIO.setup (11, GPIO.OUT)
GPIO.output (11, 1)
```

A bit more complicated, but then the Pi is a bit more complicated than the Arduino. Likewise, setting up a pin as an INPUT (with a software-based pull-up resistor, no less!) is simply done as follows:

```
GPIO.setup (11, GPIO.IN, pull_up_down = GPIO.PUD_UP)
```

If you're not familiar with the concept of a pull-up or pull-down resistor, I'll go over it when we get to switches and sensors in Chapter 10.

Continuing our trip around the board, we come to the SD card. This is your Pi's combination hard drive and RAM, so when you choose your card, give yourself room to grow. I usually recommend 16GB cards, but on the other hand, I just saw an advertisement for a 256GB SD card for around $100, so I'm torn. I'd ordinarily consider a card that size to be unnecessary, but if you're going to be dealing with video files or a lot of sensor data (as this rover may do), it might be worth the expense.

At any rate, I'd suggest getting a name-brand card, as personal experience has shown that some of the cheap generic cards are unreliable and prone to failure. At least get in the habit of backing up your card regularly, using either Linux's dd command or a similar Mac or Windows tool. It's difficult to describe the pain you feel when days and weeks of work are rendered useless by a simple SD card meltdown.

Next up: the power_in port. It's just a 5V micro USB B port (Figure 2-7), similar to the one on many cell phones or tablets. As a matter of fact, the easiest way to power your Pi is to use a standard cell phone charger. Be aware, however, that you may have mixed results, as different chargers deliver different amounts of current, and the closer you can get to 2A of current delivered, the better. If your Pi doesn't work—or acts strangely—with one charger, just try another one. You'll need at *least* 1A of power for the Pi to run without hiccups and glitches.

Figure 2-7. *Micro USB B plug*

The Pi does not have an onboard voltage regulator, so you must power it with 5 volts and only *5 volts! Those of you familiar with the Arduino are probably used to just plugging in a 9V battery and going merrily on your way. If you try that with the Pi, you'll have a nice fried, dead paperweight on your hands. If you're unsure of your charger, check its output with a voltmeter. If you're using batteries, funnel them through a regulator before sending the power to your Pi.*

The last important item on the periphery of the board is the HDMI port. Some would argue that this is where the Pi truly comes into its own, as it's capable of outputting full 1080p graphics, with 1 gigapixel/sec processing power. The onboard GPU can do Blu-ray Disc–quality playback, using OpenGL and OpenVG libraries on the chip.

That chip is located in the center of the Pi (Figure 2-8). It's a Broadcom PCM2835 system on a chip (SoC) and has an unmodified speed of 700MHz. It can be overclocked up to 1GHz if you so desire, though be aware that it *can* lead to some system stability issues. At its normal speed, it doesn't get hot enough to require cooling or a heat sink. It can be compared, performance-wise, to a Pentium III, with the graphics capabilities of a first-generation XBox. Not bad for a little computer about the size of a credit card.

Figure 2-8. *The Broadcom PCM2835*

Model B+

In July 2014, the Raspberry Pi Foundation announced the existence (and release) of the Raspberry Pi B+, an upgrade to the existing model B for the same price (Figure 2-9). It's been enthusiastically received by the Pi community. Let's take a look at how it differs from the original model B.

Figure 2-9. *The model B+*

GPIO

Probably the biggest difference in the B+ is the addition of 14 GPIO pins, for a total of 40. Luckily, the pinout of the first 26 pins remains the same, so pin connections designed to work with the first version will still work with the updated one. The additional pins give you nine more general-purpose pins, three more ground pins, and two specialized I2C ID EEPROM pins. Those two pins, numbers 27 and 28, are used to connect an I2C EEPROM, or Electrically Erasable Programmable Read-Only Memory chip. Those two pins are checked when the Pi boots up to see whether they're connected to a board; in this way, the Pi is able to detect a connected device and configure the GPIO pins to work with it. If you don't have such a board, leave those pins free, and have fun with the extra nine general-purpose pins.

USB

The B+ also has four USB ports instead of two, *possibly* making a separate USB hub unnecessary. The only thing that may be an issue is that USB 2.0 specifications state that each port should provide 5V and 500mA, for a total of 2A across all four ports; a few users have reported that the model B+ provides a total of only 1500mA across the four ports, so you may need that powered hub after all if you need to power more than three devices.

Power

The new model B+ also requires less power than its predecessor—quite a feat considering all of its extra features. The Pi Foundation replaced the old linear voltage regulators with switching regulators, which had the effect of trimming up to 1W of power from the Pi's consumption. If you're using batteries to power something like a robot, this is great news. The Pi Foundation also added a dedicated low-noise power supply for the audio circuit.

Shape

Finally, of course, the Pi Foundation moved stuff around and made the board a bit more streamlined. The foundation lined up the USB connectors with the edge of the board, combined the composite video and 3.5mm audio jack, and even added four evenly spaced mounting holes for mounting the Pi by using standoffs. There will certainly be add-on boards coming soon, designed to take advantage of the extra GPIO pins and EEPROM support, and the four mounting holes should make connecting it securely to the Pi an easy task. All in all, the model B+ is a definite step forward, Pi-wise.

Where to Get Help

It's true: as easy as the Raspberry Pi is to get started and operate, at times you're going to need help, sometimes more than a simple Google search can provide. There are a few places you should be familiar with before you go stomping off into the great unexplored expanses of the Internet, looking for answers.

The first place is on the Pi itself, and that means using Linux's `man` command. It's short for `manual`, and can be invoked for darn near any Linux command or function you're having problems with. In fact, my only beef with the `man` command is that it can be *too* comprehensive; the help pages for the simple `ls` command (which lists the contents of a directory) go on for pages (Figure 2-10), giving you more information than you ever wanted or needed to know about `ls`. Still, although `man` can be overwhelming, it provides a lot of information.

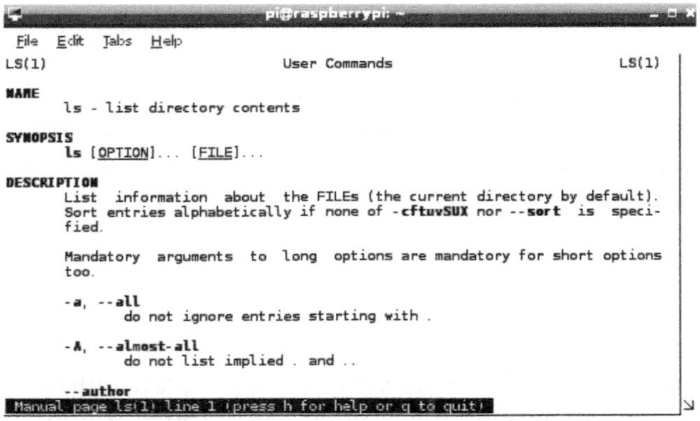

Figure 2-10. *The man(ls) pages*

The second way to get help is with Python's `help` function. Just typing `help()` at the Python prompt brings you to the interactive online help utility. When you first bring it up, it gives instructions on how to use it, like telling you to type "modules," "keywords," or "topics" if you're totally lost. Typing abs at the prompt, on the other hand, gives you the specific instructions, as you see in Figure 2-11.

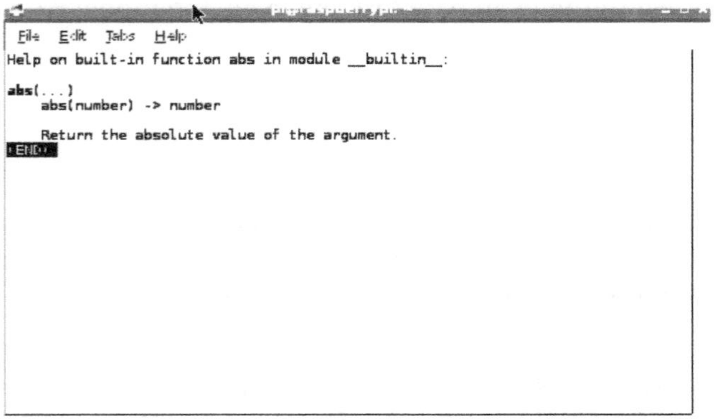

Figure 2-11. *The help() prompt*

You can go directly to the help page for a function by just including it as an argument to `help`—`help(abs)`, for instance. If you need help with a function not in Python's default libraries, you'll have to import that library first. For example, to get help with the square root function, you type:

```
import math
```

followed by:

```
help(math.sqrt)
```

If, however, neither of those onboard utilities have what you need, you'll have to ask other humans. In my experience, two sites stand out for both comprehensiveness and general helpfulness among fellow Pi programmers.

The first is the forum on the Raspberry Pi website (*http://www.raspberrypi.org/forums/*). As of this writing, there are 12,000 topics in General Discussion and 2,500 in the Python subforum. Staff and engineers from the Pi Foundation itself frequently stop by on the forum; some other members have been there since before the Pi existed in reality. There are subforums dealing with topics ranging from graphics programming and gaming—there's even a For Sale section. It's definitely worth your time to stop by and create a profile.

The other site that is well worth visiting is Raspberry Pi Stack Exchange (*http://raspberrypi.stackexchange.com*). On this free site, anybody can ask (and answer) questions—no registration is even required. It doesn't have nearly the amount of traffic as some other programming sites, but it has the advantage that it is specifically geared toward the Pi. If you're stuck, browsing its questions can sometimes be helpful.

As for other sites, Stack Overflow (*http://stackoverflow.com*)—every programmer's best friend—has many, many questions dealing with the Pi, so it's normally worth your time checking out as well. And though I knocked it at first, many times a good Google search will help. No matter how bizarre or off-the-wall your project may be, there's a good chance that somebody, somewhere has run into the same wall.

And speaking of asking and answering questions: as your experience grows, jump in and help answer questions on both sites! The Raspberry Pi community is definitely that—a community—and there is *always* someone who is just getting started who may benefit from your experience. Before you know it, you'll be helping and getting help like thousands of others!

That was a short, down-and-dirty introduction to the Pi. If you have no experience setting it up, jump to Appendix A, where I'll walk you through the steps of downloading an operating system, loading it onto the Pi, and so on. Otherwise, let's go to the next chapter, where I'll give you a quick introduction to Linux.

3 Intro to Linux

The Raspberry Pi is arguably one of the world's most popular tiny computers. It's cheap (around $35), easy to get, relatively powerful for its size and price, and easy to start using: plug in a keyboard, a mouse, and a monitor and start programming by using Python or Wolfram Mathematica or Scratch, the kid's programming language. Sometimes, you don't even need the keyboard, mouse, and monitor!

In fact, it's *so* easy to get started that many users completely forget (or ignore) that the operating system it usually runs, Raspbian, is based on Debian, a distribution of a powerful operating system, Linux (Figure 3-1). Your programs don't even have to be written in Python; the Pi's Linux roots mean it can run C, C++, Java, or even (if you're particularly masochistic) assembly code.

A World of Little Computers

Many small computers exist on the market, manufactured by companies like Parallax or Intel. They are marketed to both professionals and hobbyists, and are commonly used for things like after-market onboard car computers (*carputers*) and sensor loggers. Parallax makes an impressive eight-core microcontroller board called the Propeller, and Intel's newest entry into the market is the Galileo. The BeagleBone, by the BeagleBoard Foundation, is another popular board, and another new one on the market is Radxa's Rock, with a quad-core CPU, integrated WiFi, and an infrared receiver.

The main difference between these other, small-form-factor computers is price. Most of the Pi alternatives hover around the $100 mark, about three times as costly as the Raspberry Pi. Still, many of them are more powerful in terms of speed or computational ability, so if the Pi isn't perfect for your needs, chances are one of the others will be. For instance, the Radxa Rock sports a quad-core processor and 2GB of onboard RAM and is able to run both the Android and Ubuntu operating systems. The BeagleBone's processor, on the other hand, is comparable to the Pi's (a 720MHz ARM processor), but it also runs both Android and Ubuntu, and can be programmed with a web interface, using a version of JavaScript.

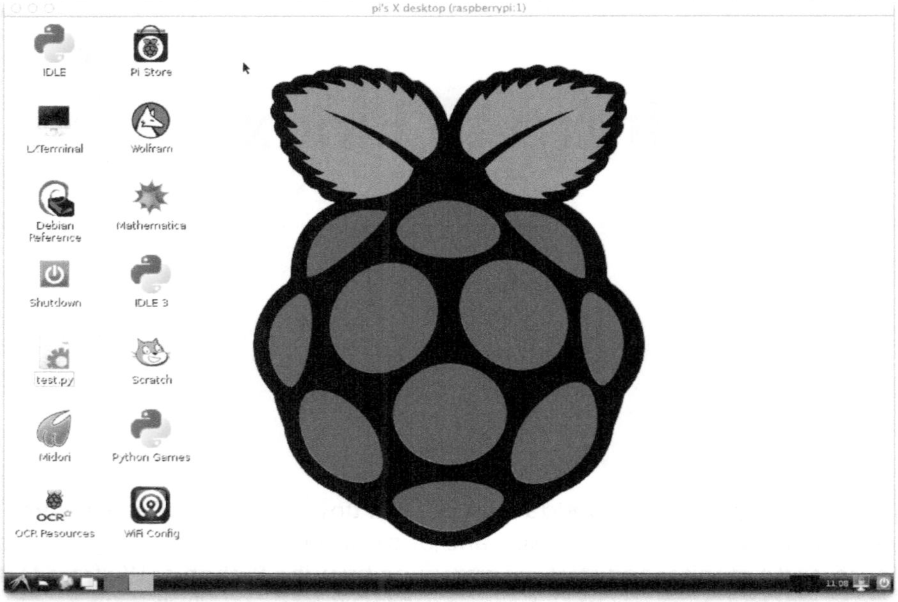

Figure 3-1. *I sense no Linux here*

Because Linux is the base of the Pi's operating system, let's do a short introduction to Linux, for those of you who'd like to know more about the OS and how to use it. Since its creation nearly 25 years ago, Linux has been considered the "geek's OS," with its proponents being viewed as the stereotypical glasses-wearing, pocket protector–having, socially inept tech crowd. In the past few years, however, Linux has become a bit more popular, and more people than you'd expect are using it. Although there doesn't seem to be one overwhelming reason for its increase in popularity, some tech experts (including yours truly) suggest that its price point (free, in most cases) and its adaptability are contributing factors. Poking around under the hood of your Windows or Mac computer can be difficult, but Linux allows you to completely rewrite, compile, and install your own kernel, making it uniquely adapted to your needs. It's also a much smaller OS, with some versions able to be installed on and run from a USB flash drive.

I can't teach you how to become a power user in a single chapter, but I can definitely make your journey through Linux Land a bit more comfortable. If you're already a Linux geek, feel free to skip this chapter, but if you're not, read on for a quick primer.

Linux was first released by its creator, Linus Torvalds, in 1991. It has its roots in Unix, BSD, MINIX, and GNU—all of them successful or unsuccessful attempts to create a completely portable operating system. Linux is written in C, and was originally intended to run on the Intel x86-based architecture. Since then, it has been ported to almost every imaginable device, from Android phones to mainframes and supercomputers to tablets to refrigerators. It's even running on computers on the International Space Station, as NASA decided it needed something more stable than Windows. In fact, Linux has been placed in so many applications and devices that it is now the most widely adopted operating system in the world.

When you interact with the Pi, you'll most likely be doing a lot of work in the terminal. With your Pi desktop up and running, double-click the LXTerminal icon to open the Pi's terminal prompt (Figure 3-2).

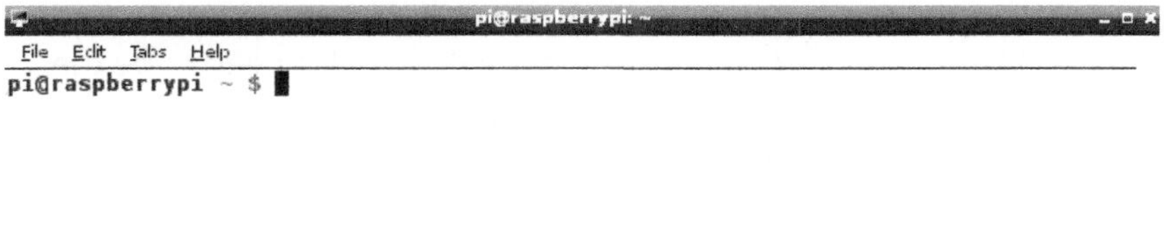

Figure 3-2. *The Pi terminal prompt*

The command line you're seeing in Figure 3-2 shows that you're the user `pi` logged into the machine `raspberrypi`, and you're in the home directory. If you were to move to a different directory, by using the `cd` command, the command line would show that location:

```
pi@raspberrypi ~/Robotics $
```

Structure

Linux, like other operating systems, is completely built around files and the filesystem. A *file* can be described as any piece of information—be it text, image, video, or otherwise—that is identifiable by a filename and a location. The location is also called the *directory path*, and keeps a file unique because technically the location is part of the filename. In other words, */home/pi/MyFiles/file.txt* is a different file than */home/pi/MyOtherFiles/file.txt*.

Filenames are case sensitive in Linux, meaning that *file.txt* is different from *File.txt*, and both are different from *FILE.TXT*. There are five categories of Linux files:

- User data files
- System data files
- Directory files, or folders
- Special files representing hardware or placeholders used by the OS
- Executable files

Each user has a default */home* directory in Linux, and within that directory you have permission to create, edit, and delete files all day long. However, if you want to edit system files, or those belonging to another user, you'll need the permissions of a special user—the *superuser*.

This superuser, also referred to as the *root user*, can edit any file in the system, including low-level system files. Because of this ability, Linux users don't log in as root unless absolutely necessary; when they do assume the root login, they log in, do what they need to, and then log out again. There's a saying among Linux users: only noobs log in as root.

There's a shortcut to gaining the powers of a superuser while still logged in as yourself: the `sudo` command. `sudo` stands for *superuser do*, and simply tells the system to execute the following command as if it were the root user issuing the command. The system will ask for the root password and then execute the command. You normally need to use `sudo` when you update or install files (hence the `sudo apt-get install` command), and when you edit configuration files like */etc/network/interfaces*. On the Pi, you also need to be the root user when you access the GPIO pins; this means that any Python program you write that accesses the pins needs to be run as root:

```
sudo python gpio-program.py
```

sudo confers great power and thus requires great responsibility. The system does not hold your hand or double-check with you when you issue a command using sudo. It's up to you to be especially certain you know what that command does before you hit Enter. It is entirely possible to completely erase the contents of your hard drive with one poorly written sudo command!

Commands

To get around in the Linux command-line interface (CLI), you use text commands to display information and run programs. You should be familiar with commands like the following:

`ls`
: List files in current directory

`cd`
: Change directory

`pwd`
: Print working directory

`rm` *`filename`*
: Remove (delete) a file

`mkdir` *`directoryname`*
: Create a directory with the name provided

`rmdir` *`directoryname`*
: Remove (delete) empty directory

`cat` *`textfile`*
: Display contents of a text file in the terminal

`mv` *`oldfile newfile`*
: Rename a file

`cp` *`oldfile newfile`*
: Copy a file

`man` *`commandname`*
: Display Linux manual of a given command

`date`
: Read system date/time

`echo`
: Echo (print) what is typed back in the terminal

`grep`
: Search program that uses regular expressions

`sudo`
: Perform as root user

`./program`
: Run a program

`exit`
: Quit terminal session

Most of these are fairly self-explanatory, but if you ever get confused, the `man` command is undoubtedly the most useful. If you are unsure of what a particular command does or what parameters or flags it uses, typing `man` *`commandname`* into your terminal brings up the manual page for that command, with more information than you'd ever want to know (Figure 3-3).

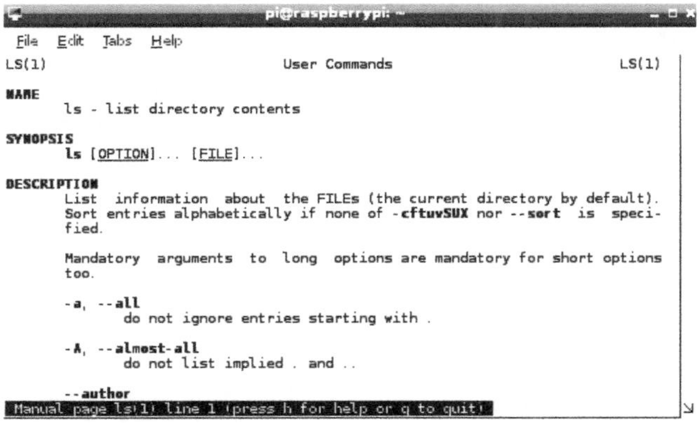

Figure 3-3. *First page of man ls*

When you first log in to the Raspberry Pi, you'll find yourself in your home directory, often illustrated in code as *~/*. Typing `ls` will show you all files and directories within that home folder, and you can then `cd` to go into other directories. If you need to move to a parent directory, you can type `cd ../` to move up one level, or `cd ../../` to move up two levels, and so on. If you get lost, typing `pwd` will tell you the full file path of the directory you're in —useful if you're several levels deep in a folder path like */home/pi/robots/test/pi/servos/test* and you've forgotten exactly how deep you are. In any case, if you are truly lost, typing `cd` with no arguments will always take you back to your home folder—a handy shortcut.

Navigation

Let's do a quick exercise to give you some practice moving around in the filesystem without using the file explorer graphical user interface, or GUI. (I'll repeat: if you're already comfortable in Linux, I won't be offended if you skip this part and move to the next chapter, where I discuss much more interesting things like rover parts.) Open your terminal, and make sure you're in your home directory by typing **`cd ~/`**.

Now make a subdirectory by typing the following command:

```
mkdir exercise
```

Then, to add a subdirectory, type:

```
mkdir exercise/subdir
```

Now you can navigate into that directory by typing:

```
cd exercise/subdir
```

Once you're in there, create a text file using the `echo` command and the greater-than (>) symbol:

```
echo "This is a test file" > file.txt
```

In the icongraphy of Linux, the greater-than symbol looks like a sideways funnel, and that is essentially its function: it *funnels* the `echo` command into a file. Now, if you list the contents of the file with `ls`, you should see *file.txt*.

Now let's rename it. You can do that by typing the following:

```
mv file.txt file2.txt
```

file.txt is now gone, replaced with *file2.txt*. Now let's copy it to one directory up:

```
cp file2.txt ../file2.txt
```

If you now move up one directory with **`cd ../`** and list the contents with **`ls`** again, you'll see two items: *file2.txt* and *subdir*. Because *subdir* still has *file2.txt* in it, you can't just do a `rmdir` on it to get rid of it. You either have to empty it first, or tell the OS to remove it *and* all files inside it. To remove the file first from where you are, you can type:

```
rm subdir/file2.txt
```

Then you'll be able to remove *subdir* with a simple `rmdir` command.

If you prefer to remove the directory and all of its contents in one fell swoop, use the `-r` flag, which will promptly delete *subdir* and all of its contents, including other directories and their files:

```
rm -r subdir
```

Use the `-r` flag with caution: it will delete *everything* in the top directory and all directories below it! We recommend staying safe and sticking with just `rmdir`, without the `-r` flag. That way, you'll be unable to delete a full directory, which may save your files from your inattention!

I'd like to conclude with a quick time-saving tip. If you press the Tab key in the middle of a long filename, the terminal will fill in the rest of the name for you as far as it can, saving you some typing. So, for example, if you need to copy *mylongfilename.txt*, type **`cp mylong`** and press Tab and the rest of the name will be filled in for you.

If you happen to have another file in the same directory called *mylongfilename2.txt*, pressing Tab will make the terminal fill in up to *mylongfilename*, and then you can finish the name with whichever one is applicable.

That is a quick-and-dirty introduction to Linux. The more you work with the Pi, the more comfortable you'll become, so don't be surprised if you find yourself switching your other computers to Linux as well.

4 Setting Up the Wireless

As you've seen, the Raspberry Pi is a full-fledged computer, if a bit smaller and underpowered compared to what the average user is accustomed to. Almost immediately after taking it out of the box, you can insert a formatted hard drive (the SD card), plug in an HDMI-capable monitor, add a USB keyboard and a mouse, and start computing. So even if your desk looks like Figure 4-1, the Pi is so small that you can power it up and start programming. It doesn't even need airflow; even when it's overclocked, the ARM processor doesn't get hot enough to need any appreciable cooling, so you can put it in a drawer or a project box and run it that way. (Yes, you can even put it in an Altoids tin, though there are a few strategic cuts you'll need to make to enable it to fit.)

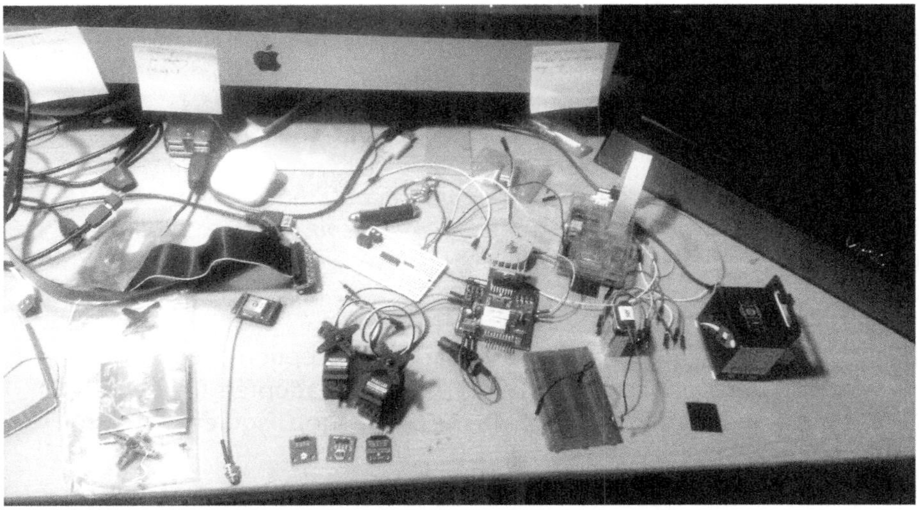

Figure 4-1. *My desk on a good day*

Once it's running, plug in an Ethernet cable and connect the Pi to your modem or router, and you can surf the Web, install updates and upgrades, change your Facebook status, and any number of other miscellaneous tasks that require Internet access, using either the built-in Midori or NetSurf browsers, or a third-party browser such as Firefox (called Iceweasel on the Pi) or even Google Chrome (called Chromium on the Pi).

But what if you don't want your Pi to be tethered to your modem or router, or even your desktop? After all, one of the huge advantages of the Pi is its size and portability; it'd be a shame to be stuck working at your desk. It'd be especially inconvenient to have a robot's brain (for example) tied to a modem, and the robot's mobility limited by the length of whatever Ethernet cable you happened to have lying around. And forget moving around outside. Robots need to be *mobile*, not tied down.

Perhaps a future version of the Pi (model C?) will come with an onboard wireless chip, even though it would probably make the Pi a bit larger. I'm not aware of any plans for a wireless model, though, which means that for the time being we've got to be proactive and add a wireless USB adapter, also called a *dongle*, to give the Pi wireless connectivity.

Historical Problems

If you've been playing with the Raspberry Pi from the beginning, you may be aware of a few problems getting it to work wirelessly. Some of those problems have stemmed from the Pi being Linux based. Other problems arise from the Pi's use of an ARM processor. (That may seem strange, as ARM processors are at the heart of millions of smartphones worldwide, but unlike the Pi, cell phones don't use a USB wireless adapter to connect to WiFi.)

> *It used to be that the mark of a true geek was someone who managed to get the wireless working on a Linux installation. I remember earning such a badge of honor on my Ubuntu 7.04 laptop, back in 2007. Wireless drivers that actually* worked *were few and far between, and many times you had to use a special tool that hacked the Windows drivers so Linux could use them. That tool, called* ndiswrapper, *is a Linux module that allows Ubuntu to use Windows drivers for some wireless cards, because the open source driver for the card either doesn't exist or doesn't work for whatever reason.*

Times have changed, of course, and (with a few unremarkable exceptions) Linux's wireless problems have become virtually nonexistent. That is, until early adopters fired up the Pi, plugged in a USB dongle, and tried to get online in the coffee shop (Figure 4-2).

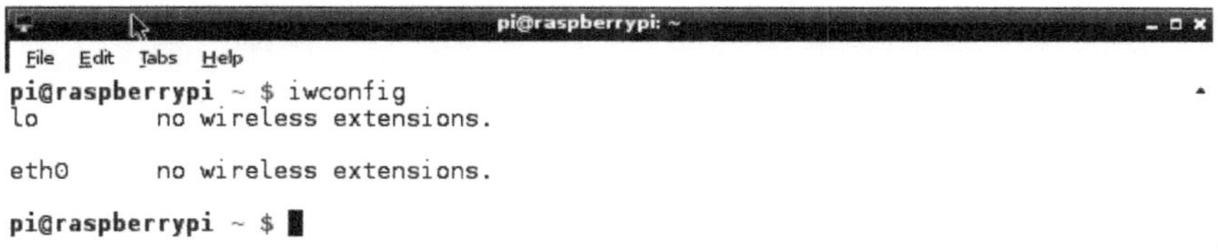

Figure 4-2. *No happiness in Wireless Town*

If the Pi recognized the wireless adapter at all, it seemed a miracle. Most of the time the adapter didn't fire up, but if it did, the Pi didn't recognize it as a network device. Linux drivers may or may not have existed for different devices, but even if they did, they didn't work on Raspbian for whatever reason. Raspberry Pi forums filled with the lonely cries of stranded users: "How do I get wireless working on my Pi?" Luckily, the Pi community worked together and determined that all you need is the right adapter.

The Ralink Chipset

As it turns out, only wireless adapters with certain chipsets play nicely with the Pi, something that many of us found out the hard way. Chances are extremely good that the D-Link or Netgear adapter you picked up from the big box store will run sketchily on the Pi, if it runs at all. Sometimes you may get lucky; if the chipset is compatible and the adapter doesn't draw too much current, you're golden. If, however, the adapter draws too much current, you'll most likely experience other problems, such as sporadic mouse and keyboard connectivity problems, and even intermittent shutdowns. For that reason alone, it's always a good idea to plug your wireless adapter directly into your Pi's USB port, and plug the mouse and keyboard—if you're using them—into a USB hub, assuming that you're using a *nonpowered* hub. If you use a powered USB hub, you can plug the adapter into it without problems. Otherwise, the Pi may not send enough current through the hub to power the adapter.

The thriving Raspberry Pi community has experimented with any number of adapters, saving us from having to go through them all one by one. As it turns out, those dongles with the Ralink RTL8188CUS chipset will work almost without exception. The Pi community has found that the following two adapters work excellently:

- Edimax EW-7811UN (Figure 4-3)
- Ralink RT5370 (Figure 4-4)

Figure 4-3. *Edimax EW-7811UN*

Figure 4-4. *Ralink RT5370*

Both of these are available on Amazon and eBay for reasonable prices (less than $10). In fact, I suggest you order more than one; they're so small that I've actually lost a few, and when (not *if*) you order another Pi, you can equip it as soon as it arrives. If you choose to go off on your own and explore the compatibility of a different adapter and/or chipset, and if you have some luck getting it to work, you're encouraged to post your findings on the Raspberry Pi forums (*http://www.raspberrypi.org/forums/*).

Making It Work: The GUI Way

When you get your adapter, plug it into the Pi. On your Pi's desktop, you'll see an icon called WiFi Config (Figure 4-5). Double-click it, and you'll see the configuration screen (Figure 4-6).

Figure 4-5. *The WiFi Config icon*

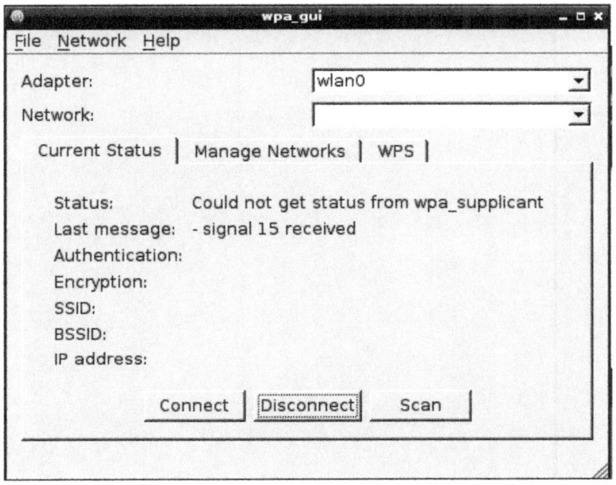

Figure 4-6. *The Wifi Config screen*

The adapter, `wlan0`, should show up in the top drop-down list, and your wireless network should show up below that. With your network selected, click the Manage Networks tab. Select your network and click Edit, which will bring up the NetworkConfig window (Figure 4-7).

Enter your network information and click Save. Now just go back to the Current Status tab and click the Connect button. The adapter should negotiate a lease from your DHCP-enabled router and hop onto your wireless network. To make sure it worked, in your terminal window, type **`ifconfig`**. You should see three Ethernet adapters listed: `eth0`, `lo`, and `wlan0` (the newest one). Assuming you're still connected via Ethernet cable, your `eth0` adapter should have an `inet` address listed, and now your `wlan0` adapter should have one as well. Voilà! Your wireless is now working!

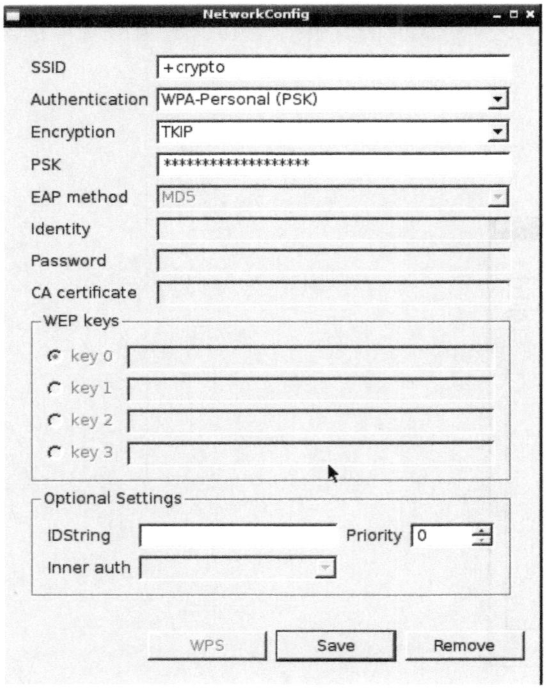

Figure 4-7. *NetworkConfig window*

Making It Work: The Command-Line Way

On the other hand, perhaps this whole process didn't work. Or perhaps you're feeling nostalgic for the old days, when you had to wire up your adapter (no pun intended) with bubble gum and baling wire. So let's look at the process of setting up a wireless adapter using the command line and editing some files.

It's pretty easy, and requires only one package and one file edit. First, in your terminal, update and upgrade your Pi:

```
sudo apt-get update
sudo apt-get upgrade
```

This is *always* a good idea before a computing session when you're going to be installing new packages and software. Many Linux distros notify you of available updates; Raspbian, however, does not, so checking once a day is a good habit to get into.

Once you're upgraded, try the following:

```
sudo apt-get install wpasupplicant
```

You'll most likely get a message stating that *wpasupplicant* is already the newest version; I believe it's included in all newer releases of Raspbian. In case you have an old version, however, you'll need to download and install it.

Once it's installed, use a text editor to open its configuration file:

```
sudo nano /etc/wpa_supplicant/
wpa_supplicant.conf
```

Here is where you put in your network information. (Coincidentally, it's where the information is stored if you enter it via the NetworkConfig GUI in the previous section.) Leave the first lines of the file as they are, but add the following lines:

```
network={
   ssid="_your network id_"
   psk="_your network key_"
   proto=WPA
   key_mgmt=WPA-PSK
   pairwise=TKIP
   auth_alg=OPEN
}
```

Replace *your network id* and *your network key* with your network name and key. (Your network key is the password you've set to log into your network from your devices.) It's true, you're putting your network password into a configuration file, and it's in plain text. For those of you with the tinfoil hats, this means that if your Pi falls into your arch-enemy's hands, that person will have the key to your network. Unfortunately, there's no way around this. It *is* a configuration file, after all, and the Pi's operating system must be able to "read" it. You can only hope that any nemesis who gets their hands on your rover is not Linux-savvy.

You may have to fool with the settings to match your network configuration; some older versions of *wpasupplicant* won't work with WEP authentication. Think of it as a kick in the pants to change your security settings if you haven't already; WEP is *not* a secure standard.

Save the file and reboot your Pi:

```
sudo shutdown -r now
```

When your Pi comes back up, your wireless adapter should connect to your WiFi router.

Setting a Static IP Address

Getting your wireless working, although impressive, is only half the battle. You'll also need to set a static IP for your adapter. If you don't, it's entirely possible that every time you power on your Pi, it'll receive a different IP address from your router. Because you need to log in remotely in order to control your robot (for those times you're controlling it on your home network), you're going to need to know the IP address it's using. When you set a static IP, the router will always assign the same numeric sequence to your Pi's wireless adapter.

This part is easy. In your terminal, open your *interfaces* file:

```
sudo nano /etc/network/interfaces
```

If this is the first time you've opened the file, it's probably going to look something like this:

```
auto lo

iface lo inet loopback
iface eth0 inet dhcp

allow-hotplug wlan0
iface wlan0 manual
wpa-roam /etc/wpa_supplicant/wpa_supplicant.conf
iface default inet dhcp
```

The first part of the file is fine, but you're going to edit the rest of it to connect to your network without using DHCP. Just a warning: again, this requires putting your network SSID and password in plain text in your *interfaces* file, so try to keep your Pi out of your enemies' hands.

Edit the file so it looks like this (I'm using my own address, netmask, and other values—substitute your own):

```
auto lo

iface lo inet loopback
iface eth0 inet dhcp

auto wlan0
allow-hotplug wlan0
iface wlan0 inet static
address 192.168.2.60
netmask 255.255.255.0
broadcast 192.168.2.255
gateway 192.168.2.1
dns-nameservers 8.8.8.8 8.8.4.4
wpa-passphrase my-passphrase
wpa-ssid my-ssid
```

The `address` is the IP address you'd like to assign to your Pi. The `netmask` establishes the subnet you're using; unless you have a huge subnet in your house, chances are yours will also be `255.255.255.0`. The `broadcast` value will mirror the first three values of your subnet, with a `.255` at the end, and the `gateway` is the IP address of the router on your network.

You'll notice you don't need the `wpa_supplicant` and `iface default` lines; you can either remove them or comment them out by placing a # at the beginning of the line. When you've edited the file, save it, and reboot your Pi with the following:

```
sudo shutdown -r now
```

When your Pi comes back up, you should be connected to your network with full Internet access and a static IP. To test it, open a terminal and first type **`ifconfig`** and make sure that the address listed for `wlan0` is the one you chose and entered in the *interfaces* file. Finally, make sure you can access the network by typing **`ping google.com`** and see that you get some responses.

Now that your Pi has a static IP, and because you enabled the SSH server when you set up your Pi with the `raspi-config` tool (see Appendix A if you need help with this), you can now log into your Pi with this command, giving it the password `raspberry` when you're prompted:

```
ssh -l pi 192.168.2.60
```

If for some reason this fails, fear not—all is not lost. You will, however, need to connect physically to your Pi with a keyboard/mouse/monitor combination to proceed. (For what it's worth, I *always* physically connect to my Pi this way when I'm screwing with networking stuff. That way, I still have access to it, even if I completely kill my network adapter.)

When you have the Pi hooked up to a monitor, open a terminal and type **`ifconfig`** to see exactly what IP addresses you *do* have. You should have three adapters listed: `eth0`, `lo`, and `wlan0`. If the adapter is working and active, it should have an `inet addr:` listed in the second line of its description. If there is no address listed, and you've followed all of the preceding steps, you may have a defective wireless adapter. Try another dongle (preferably one with the chipset mentioned earlier) and see if you have better luck.

Network Addresses and Terminology

If you find yourself confused by the various addresses and netmasks and whatnot, don't worry too much. The terminology and concepts can get tricky, as they are used to define and control various portions of the Ethernet networking protocol standards defined by the Institute of Electrical and Electronics Engineers (IEEE). Certain IP addresses are reserved for private networks (e.g., `192.`*`xxx.xxx.xxx`* and `10.`*`xxx.xxx.xxx`*). The `netmask` defines what addresses are available within a certain subnet (e.g., `192.168.2.`*`xxx`*), and the `broadcast` and `gateway` addresses are usually set by (and configured with) your router. The `gateway` address, for instance, is normally your router's IP address on the network, and *usually* ends in `.1`. The `dns-nameservers` I use are Google's servers; you may have others supplied by your ISP.

If that weren't complicated enough, soon devices will be moving to the IPv6 standard, which will define 2^{128} IP addresses. Instead of a `192.168.2.1` format, IPv6 addresses are in hex, such as `7c:6d:62:73:b3:84`. When that happens, we'll gain more addresses than we could ever use, but it *will* have the effect of making the */etc/network/interfaces* file a bit more complex. Stay tuned.

Running the Pi Headless

No, this doesn't mean you're decapitating your Pi. A *headless* configuration refers to the practice of doing your Pi-related work over a network connection, rather than attaching it to a monitor, keyboard, and mouse. I do all of my Pi work this way, as do many other users. With the wireless adapter working and a static IP address configured, you can store the Pi anywhere in your house or office, and simply log into it to work on it. It's also how you'll be working with it after it's set up on the rover, as you're not going to be following the rover around with a keyboard and a monitor.

The main protocol you'll be using to log into your Pi remotely is Secure Shell, or SSH. If you're using a Mac or a Linux machine as your desktop computer, you won't need any special tools, as those machines have SSH clients built in. The Pi has a built-in SSH server, which can be activated (if you didn't already) using the `raspi-config` tool (see Appendix A for more information). If you're using Windows, you'll need a client called PuTTY (*http://www.putty.org*). No installation is necessary; just download the program and place it on your desktop or wherever you'd like.

When you're ready to log in, just open an SSH connection to your Pi's IP address. On a Mac or Linux machine, the syntax is as follows:

```
ssh -l pi 192.168.2.60
```

(Using whatever your Pi's IP address is, and assuming you haven't changed the default username from `pi`.) Then just enter the password when prompted, and you'll be working on the Pi via the command line, as if you had a terminal open in a standard desktop session. On a Windows machine, just open PuTTY, leave everything as the default, put the Pi's IP into the Host Name (or IP Address) field, and click Open.

If you would like to see a desktop interface but still want to work remotely, you can configure the Pi as a Virtual Network Computing (VNC) server. Once the server is installed, you can start it on the Pi, start up a VNC client on your desktop machine, and again work remotely, only with a full desktop. The easiest package to get running (in my opinion) is TightVNC. Open a terminal on the Pi and install it with the following:

```
sudo apt-get install tightvncserver
```

Once it's installed, you can start the server with a command line:

```
vncserver :1 -geometry 1024x768 -depth 16
```

Now all you need to do is install a VNC client on your desktop machine (I use Chicken of the VNC (*http://sourceforge.net/projects/chicken/*) on my Mac) and connect to the Pi's VNC server.

Setting Up an Ad Hoc Network

The last thing I'd like to go over in this chapter is setting up your Pi to form an ad hoc network. Although being wirelessly connected to a network is great if you're at home or school or the office, you won't be able to connect to your rover if there's no local network. And because rovers do best, well…*roving*, you'll probably want to take it outside.

The solution here is to set your Pi to become a server of an ad hoc network. With its own DHCP service, it can hand out DHCP leases to your laptop or your phone, allowing you to `ssh` directly into it, even outdoors. (An *ad hoc network* is simply a small, computer-to-computer network. If you were to connect your Pi directly to your laptop with an Ethernet cable, you would have a small, wired, ad hoc network.)

You'll need to edit your *interfaces* file again with this command:

```
sudo nano /etc/network/interfaces
```

You may want to save the current file as *interfaces-old* or something similar before you start poking around. I sometimes switch back and forth between *interfaces* files, depending on my plans for the Pi that day. Sometimes it's just easier that way.

When it's saved, edit the wireless portion of the file so it looks like the following:

```
auto wlan0
iface wlan0 inet static
address 192.168.1.1
netmask 255.255.255.0
wireless-channel 1
wireless-essid RPiWireless
wireless-mode ad-hoc
```

Save it and exit. Then restart your wireless connection with the following:

```
sudo ifdown wlan0
sudo ifup wlan0
```

Now you should be able to see a network called RPiWireless from any other computer with a wireless connection.

Don't try to connect to it yet, however. Your Raspberry Pi has created an ad hoc network, and is broadcasting it, but that's all your Pi is doing. Any other computers trying to connect to that network will never be assigned an IP address, and thus will never be able to connect to it and communicate with the Pi. We need to set up the Pi to assign IP addresses to connecting computers, and we'll do that with a DHCP package called, appropriately, `isc-dhcp-server`. First, update your Pi and then grab the package:

```
sudo apt-get update
sudo apt-get install isc-dhcp-server
```

When it's finished installing, you need to configure it using its configuration file. Open it with this command:

```
sudo nano /etc/dhcp/dhcpd.conf
```

There's a lot of preconfigured information in that file. In general, you'll want the only uncommented lines to be like this:

```
ddns-update-style interim;
default-lease-time 600;
max-lease-time 7200;
authoritative;
log-facility local7;
subnet 192.168.1.0 netmask 255.255.255.0 {
 range 192.168.1.5 192.168.1.100;
}
```

Now restart your Pi and look for the wireless network again with your other computer. Not only should the network show up, but you should be able to connect to it. Once you're connected to the Pi's ad hoc network, you can ssh into your Pi by typing **`ssh -l pi 192.168.1.1`**. This is the command you'll use to connect to your rover after you take it outside, away from any WiFi networks. Moreover, with the DHCP server running, you won't have to set up your laptop with a static IP address; the Pi will assign you one every time you connect to the network. You *can* assign your laptop a static IP if you wish, but it's unnecessary, particularly because that IP address will be valid only during the times you're logged into the Pi's ad hoc network. Otherwise, you'll have to reconfigure your laptop's IP address every time you log on and off the Pi's network.

Now that we have the wireless working, let's look at building the rover.

Parts Is Parts 5

OK, enough introductions and tours and histories and whatnot; you're probably itching to get to the building of the rover. With that in mind, let's take a look at the parts used to build it. I've included a section just for sensors; remember that you can add or take away from these as you like. The more sensors you pack on board, however, the more like a rover your machine becomes. I've also, where applicable or possible, added the links to where you can buy these parts online.

Remember: you don't *have* to use the parts I list here. For instance, I'm using the wheels from a Power Wheels Escalade, but you may decide to use larger or smaller tires (especially if you make your rover's body quite a bit smaller than mine).

For that matter, you can change the design to make a smaller rover, which has its advantages: a smaller rover needs less wood, so it will be lighter in weight, which probably means you can use smaller motors, and so on. That being said, there's not much as impressive as a big beefy rover driving around, putting all of the RC cars in the neighborhood to shame. Plus, if you want a lot of sensors on your rover, you're going to need more space to put them...

Body

The following is a list of materials you may end up using for the body of the rover:

Wood (1 × 4), 6′
: Used for the walls of the rover body. Use something lighter in weight like pine or even balsa. Oak or mahogany are probably not good choices.

Plywood, 4′ × 4′
: Used for the bottom of the rover. Again, lighter is better.

Plexiglass, 2′ × 4′
: This is totally optional. I wanted to put a lid over the rover's contents, but thought it would be cool if you could still see the guts. It's purely for aesthetics, as the thickness I used (1/16″)

is much too thin to offer any structural protection. If you live in a particularly warm or sunny climate, you may find that the plexiglass gives a sort of greenhouse effect to your rover, with the unwanted side effect of heating up your electronics; experiment and use with caution, perhaps with a few holes cut for venting, along with the cuts necessary to mount the robotic arm.

Aluminum channel (http://bit.ly/1sZYEXs), 2'
Used for the robotic camera arm (Figure 5-1). Also called *C-channel*, this is often used to surround glass or plexiglass in building projects. Get something as light as you can, yet beefy enough to hold the camera(s) and run the cables through.

Figure 5-1. *Aluminum C-channel*

Small hinges (2)
Used to attach the plexiglass lid to the main body.

Angle brackets (2)
Used to mount the motors to the body of the rover. This may take some creativity on your part, as every rover motor and body will differ slightly. After much thought, I ended up using the brackets shown in Figure 5-2, with some modifications that you'll see in Chapter 7.

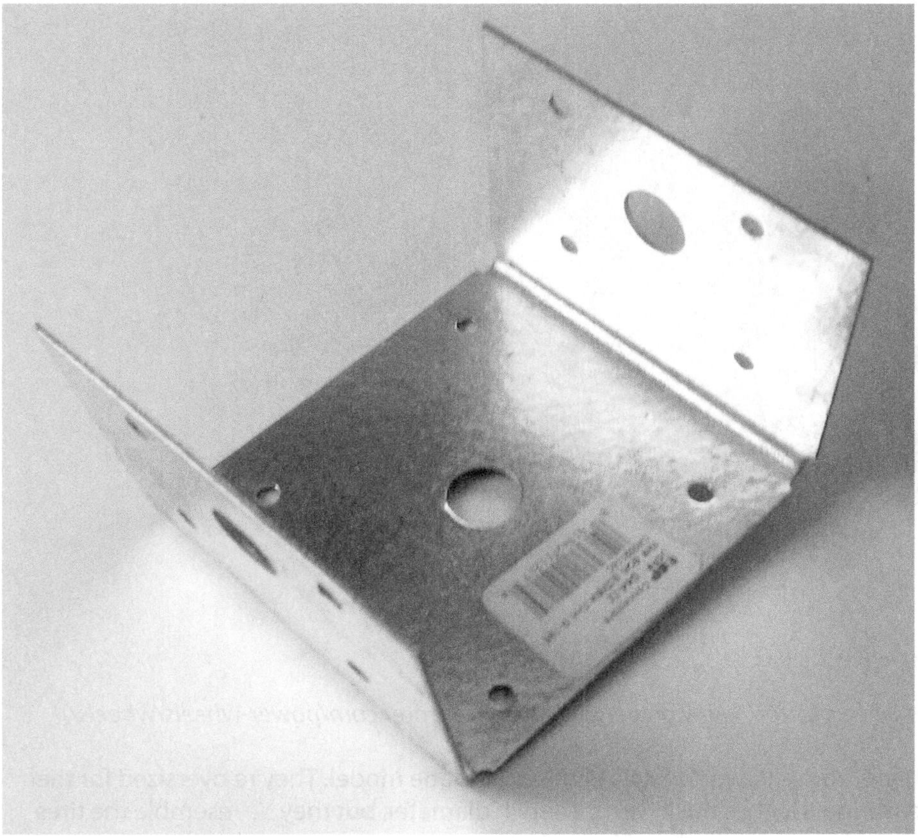

Figure 5-2. *Motor-mounting bracket*

Wheels, Motors, and Power

The following is a list of materials you'll need to get your rover powered up and moving:

Motors (http://www.amazon.com/gp/product/B005IR1NBA/) (2)
: I picked these up on Amazon for about $25. They are designed to move electric seats in a car, and the long motor shafts make them ideal for attaching to a wheel (Figure 5-3) or an axle shaft.

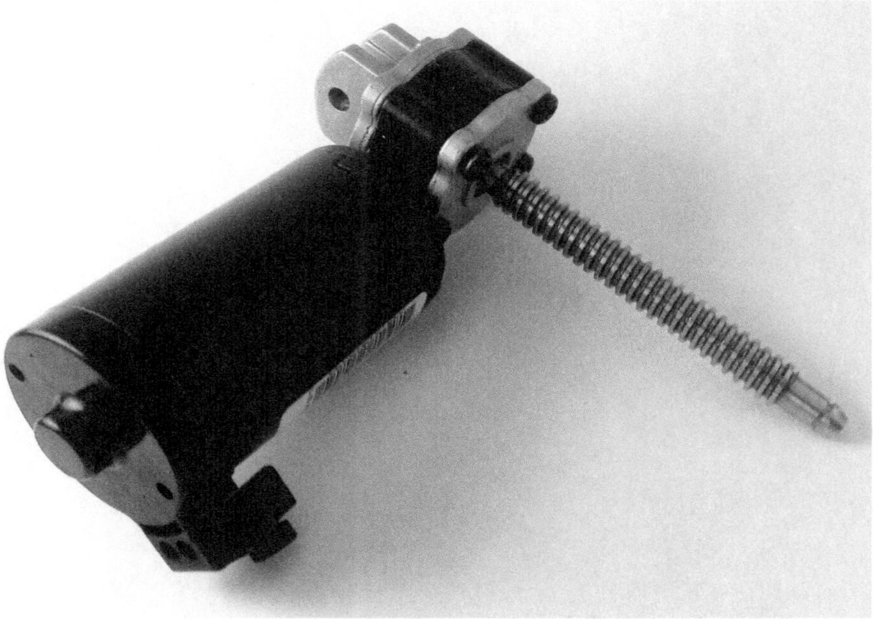

Figure 5-3. *Seat motor*

Power Wheels wheels (http://www.powerwheelsservicecenter.com/power-wheels/wheels/) (4)

I used wheels from a Power Wheels Cadillac Escalade model. They're oversized for the rover, measuring 8 inches thick with a 14-inch diameter, but they *do* resemble the tires on NASA's rovers, so there's that (Figure 5-4).

Figure 5-4. *Escalade wheels*

Aluminum axle (http://bit.ly/1s3TTLi)

From any hardware store. I used stock with 1/2" diameter, in order to fit the wheels (Figure 5-5). You'll need only one of these, for the front wheels. The rear wheels are direct-powered by the motors.

Figure 5-5. *Aluminum axle*

Bearings (http://bit.ly/1s3U89b) (2)
: Again, from a hardware store, with a 1/2″ inner diameter to fit around the axle and small enough to fit inside the hub of the wheel.

High-torque servo (http://www.pololu.com/product/1057)
: Used for the robotic camera arm. The high-torque version is necessary because the arm is pretty heavy, even if it *is* made of aluminum (Figure 5-6).

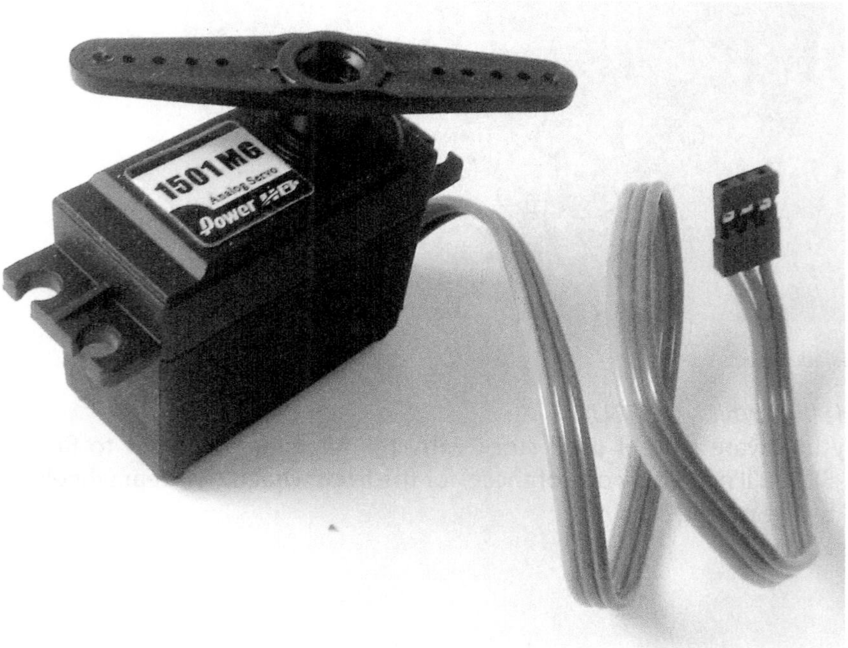

Figure 5-6. *High-torque servo*

12V battery

Purchased from a battery supply store. Get one that's used for things like electric scooters and wheelchairs. It should be large enough to power your motors for a decent amount of time (Figure 5-7). If you have the means, get two batteries, as more power is *never* a bad thing.

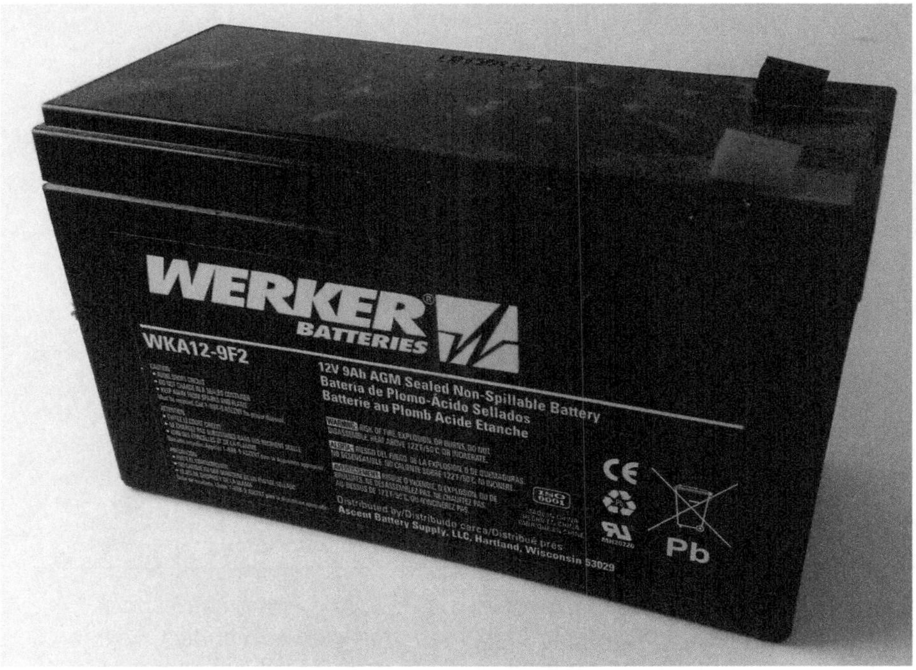

Figure 5-7. *12V 9Ah battery*

Li-Poly RC battery (http://bit.ly/1rdAuRE)

Used for powering the Pi. You don't need an especially powerful or big one. I use 1.3 milliampere-hour (mAh) battery packs, which are incredibly small and light and last for well over an hour on a single charge (Figure 5-8).

Don't forget to purchase a charger that's compatible with whatever battery you choose. Keep the voltage under 12 volts, as the voltage regulator we'll be using (the Pi has no onboard regulator) can safely handle that much, but not less than 5 volts. 1.3 to 1.8 mAh should be plenty, as the Pi draws about 1A.

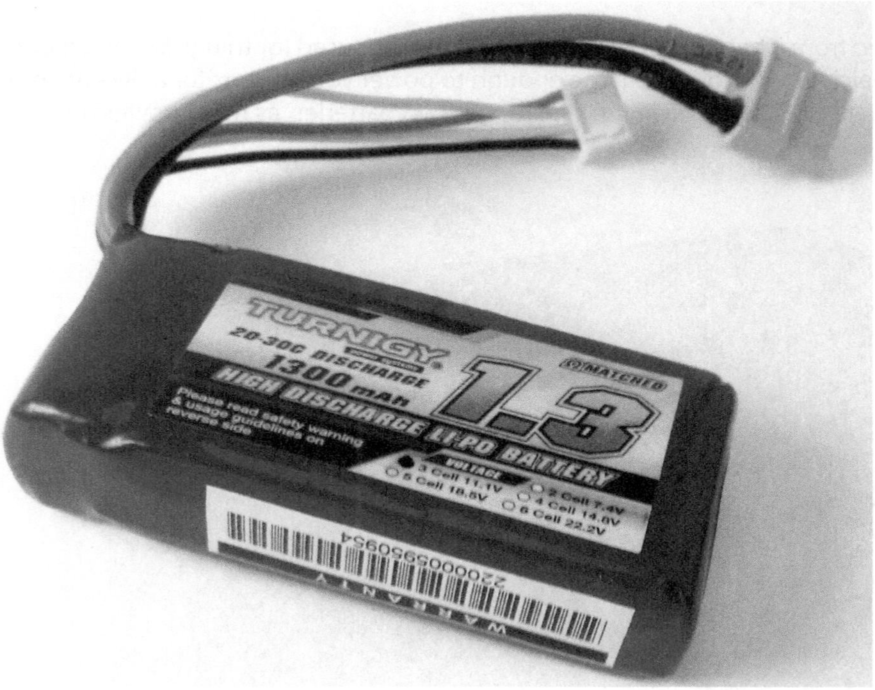

Figure 5-8. *Li-Poly battery*

The lithium polymer (Li-Poly) batteries used for remote-control vehicles pack a lot of power into a small package. Be extra careful when you connect them to make sure you don't short out the leads. Li-Polys have been known to burst (some say explode) or catch fire when shorted. That could put a big dent in your rover-building plans.

USB car charger (http://amzn.to/Zra3C2)
: These are nice because they have a built-in voltage regulator that can bring the incoming voltage down to the Pi's required 5 volts (Figure 5-9). To this, add a short USB cable (USB type A to USB micro).

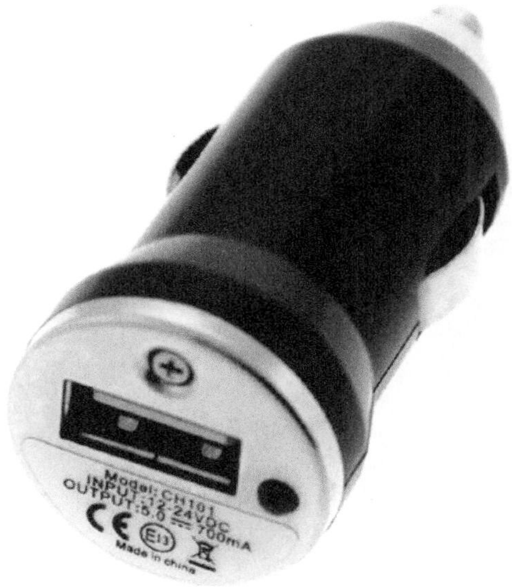

Figure 5-9. *USB car charger*

Female cigarette lighter socket (http://amzn.to/ZracoW)
: To this, add a few connectors for the type of Li-Poly battery that you use (XT60, for instance). They should be available at the same place you bought your battery pack. I'll show you how to connect everything to give yourself a quick-connect setup in Chapter 7.

Sensors

Some of these sensors are bought individually, such as the GPS unit and the SHT15 thermometer. However, you can also order a 37-in-1 sensor pack from a company called Deal Extreme (*http://bit.ly/1BJ8dsi*) for under $40. This great deal includes a temperature sensor, IR sensor, rotary encoder, Hall effect sensor, and on and on. You can pick and choose which ones you'd like to add to your rover, and you're limited only by room and power requirements.

At the bare minimum, I suggest the following:

GPS unit (https://www.adafruit.com/products/746)
: See Figure 5-10.

Figure 5-10. *GPS unit*

SHT15 temperature sensor (https://www.sparkfun.com/products/8257)
See Figure 5-11.

BMP180 barometric pressure sensor (https://www.adafruit.com/products/1603)
See Figure 5-11.

HMC5883L magnetic field sensor (https://www.adafruit.com/product/1746)
See Figure 5-11. This sensor, similar to a compass, can tell you how it is oriented with respect to the Earth's magnetic field. It's especially nice because it can be programmed to read in three dimensions, rather than just two.

Figure 5-11. *Thermometer, barometer, and magnetometer*

Accelerometer (https://www.adafruit.com/products/1231)
: This sensor is similar to what you find in your smartphone. It detects acceleration in any direction, including that due to gravity, which means it can detect which way is "down" and let you know if your rover is on a slope, or flat ground, or upside-down.

Webcam
: This may require some experimentation on your part before adding it to the rover, as so many brands and models of webcam are out there, and some work better than others on the Pi. At home, when my Pi is plugged into main power, I use a Logitech C510 webcam because it has pretty good image quality. In the field, when I'm running the Pi on batteries, I have a battered old webcam of unknown origin. (Honestly—I've had it forever, and the make and manufacturer are worn off the casing. It barely draws any power, however, so I keep using it.)

Photoresistor
: This is a resistor that changes its amount of resistance based on the amount of ambient light. As the amount of light increases, the resistance decreases. It's a handy way of making a device light-activated, for instance, or determining whether your rover is in the shade. They're available from any electronics store, including RadioShack, if you happen to have one nearby.

Ultrasonic rangefinder (http://amzn.to/1rSdTxQ)
: This handy little device, shown in Figure 5-12, uses ultra-high frequency sound waves to determine the distance to an object in front of it. It is useful for determining whether your rover is about to run into a wall.

Figure 5-12. *Ultrasonic rangefinder*

Magnetic field sensor (http://secure.robotshop.com/en/phidgets-magnetic-sensor.html) (Hall effect sensor)
: This device senses changes in the surrounding magnetic field, and is often used to detect movement between pieces of metal or on a sliding or rotating sensor platform.

Infrared motion sensor (http://bit.ly/1vIQH6R)
: This sensor detects motion by sensing changes in the surrounding IR field.

Miscellany

The following lists other miscellaneous parts you may need during construction:

Breadboard
: If you've done any electronics work, you probably have at least one of these floating around your workshop (Figure 5-13). They're great for connecting parts without soldering, allowing you to move and redesign your circuit as many times as you like. They also make it easy to have a common power and common ground line, and if you have several sensors that use the I2C protocol (Chapter 10), it's a handy way of connecting them all to the I2C bus.

Jumper wires
: Used to connect parts on your breadboard. I recommend checking eBay for a pack of wires with male-male, male-female, and female-female ends.

MCP3008 analog-to-digital chip (https://www.adafruit.com/products/856)
: This chip is used to convert the analog signals received from some sensors, like the photoresistor, to digital signals that the Pi can understand.

Edimax EW-7811UN wireless adapter (http://amzn.to/1vIRl4k)
: Or a similar one with a compatible chipset—see Chapter 6.

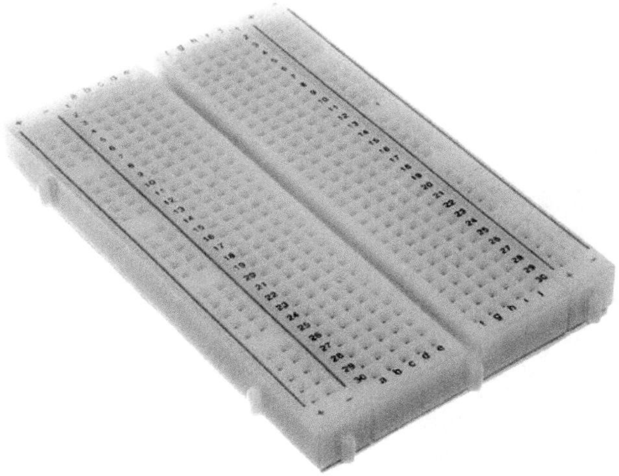

Figure 5-13. *Breadboard*

Dual H-Bridge L298H motor controller (https://www.sparkfun.com/products/9670)
: This handy motor-controller board, shown in Figure 5-14, can control two motors using a power source completely separate from the Pi. In other words, you can power your Pi with a small 1.3mAh battery pack, and simultaneously power your motors with a beefy 12V 9Ah battery, all without worrying about burning out your Pi by channeling too much current through it. It's based on the common L298 chip, and I'll go over how to use it in Chapter 8.

Common workbench supplies
: Glue, paint, screws, etc.

Tools

In addition to all of these items, you'll need some standard tools:

- Dremel multitool
- Soldering iron (I use a Weller, but all I recommend is that you get a high-quality iron that allows you to adjust the temperature of the tip. Your projects will thank you.)
- Hacksaw
- Cordless drill
- Pliers
- Wire cutters
- Screwdrivers

Figure 5-14. *L298H dual motor controller*

A good workspace, where you can spread out and lose things/get organized, is invaluable. If you can't get a large work area to call your own, an understanding spouse/roommate who will let you use a common space is the next best thing. Just remember to clean up afterward.

6 Installing ServoBlaster

The current design of the rover has only one servo: a high-torque model used for raising and lowering the robotic arm (Figure 6-1).

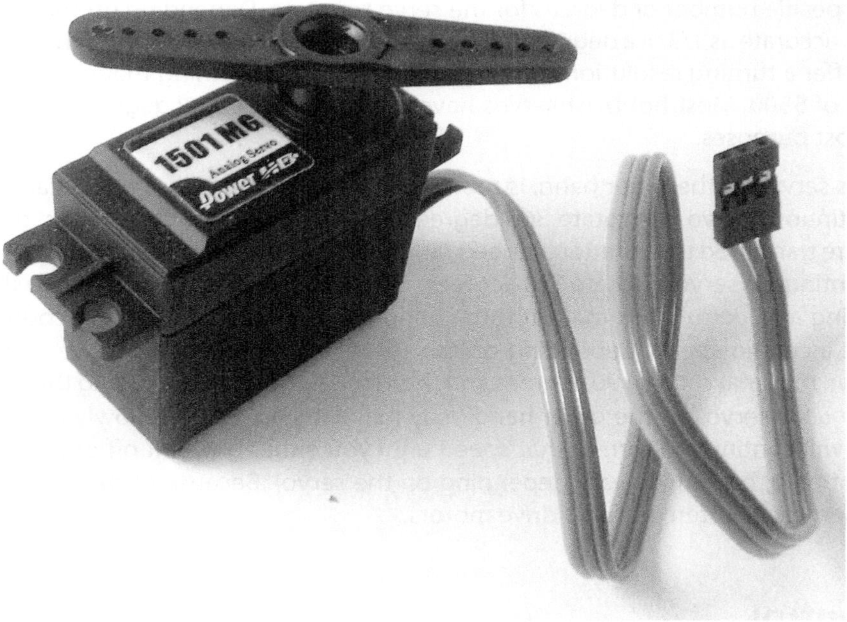

Figure 6-1. *High-torque servo*

However, you may decide to add more servos as the build progresses. You could change the design of the arm to give it two or three degrees of freedom (DOF), or you might add other attachments such as a gripper claw, a sample return arm, or a laser turret. Whether you keep the rover simple, with just a robotic arm, or you add to its capabilities, you'll need to be able to control those servos smoothly and effortlessly.

Servos

At its core, a *servo* (or *servomotor*) is really nothing more than a DC motor, over which you have a certain amount of fine-grained control. How *much* control depends on the type and make of servo and the software or hardware you're using to control its movement. Probably the most common place to find servos is in radio-controlled vehicles, but they're also used in animatronics, robotics, and other types of automation. They are often mechanically linked to a small potentiometer, which sends positional feedback to the controlling software or hardware. In this way, the controller is always aware of the servo's position, and can send or modify commands based on that position.

There are two types of servos: *standard* and *continuous*. The core of the servo remains the same; the only difference is what you can do with it. A standard servo can rotate between 0 and 180 degrees (a *half turn*) or only between 0 and 270 degrees (a *three-quarters turn*). When you send a servo a command to rotate to a specific position, that command is translated into a specific number of degrees for the servo to rotate. Depending on the servo, this can be as accurate as 1/3 of a degree. There are even some high-precision, high-quality servos that offer a turning resolution of 0.07 degree, though something that precise can cost upward of $500. Most hobbyist servos have a precision of a few degrees, which is plenty for most purposes.

A continuous servo, on the other hand, is more like a regular DC motor than a standard servo. A continuous servo can rotate 360 degrees; when you send it commands, those commands are translated to a rotational speed rather than a position. This means that you can't tell a continuous servo to rotate 720 degrees (two complete rotations) and then stop. Rather, sending a value to it will make it turn continuously, either clockwise or counterclockwise, at a certain speed, depending on the value sent. Sending the value `75` to a standard servo may make it turn 90 degrees and then stop, for example. Sending the value `75` to a *continuous* servo, on the other hand, may make it start turning slowly counterclockwise. It will continue to turn at that speed until you either power it off or send it a stop value (often `0`, but it may vary depending on the servo). Because of this behavior, continuous servos are often useful as drive motors.

PWM Control

You don't *need* a special library to control the servos. As it turns out, the *RPi.GPIO* library we use to interact with the Pi's GPIO pins is capable of the type of output necessary to control servos: pulse-width modulation, or PWM.

The RPi.GPIO library is included in all recent Raspbian distributions, though not in the early versions. To see whether you have it, fire up a Python prompt on your Pi with the command:

```
python
```

Then type the following command:

```
import RPi.GPIO
```

If you get an `ImportError`, you can easily install the module by exiting your Python prompt with `Ctrl-D` and then type the following:

```
sudo apt-get install RPi.GPIO
```

If it still doesn't work, you might have to completely delete your current RPi.GPIO and install a new one afresh:

```
sudo apt-get remove RPi.GPIO
sudo apt-get install RPi.GPIO
```

PWM is simply the process of sending precisely timed bursts of power to a receiver, such as an LED or a servo. To be understood and acted upon by the servo, the pulses must be at a particular frequency—most often 50Hz, or one pulse every 20ms. It is the width—or length—of the pulse that determines what the servo will do. An ON pulse that lasts 0.5ms, sent every 20ms, will send the servo all the way to the left, or counterclockwise. A pulse of 2.5ms, on the other hand, will send it all the way to the right, or clockwise. Different pulse widths between those two extremes will send the servo to any position you choose.

So how do you send these millisecond pulses to the Pi accurately? Like all processes running on the Pi at any given time, Python is constantly being interrupted (even if only for a millisecond or two) by system-level processes. This makes precise timing of a pulse impractical, if not impossible. Luckily, the *RPi.GPIO* library has what we need: the ability to set a particular pin as a PWM pin and give it a duty cycle necessary to power a servo.

When I say "duty cycle," I'm referring to the length of time the signal is HIGH during each 20ms span of time. It's probably best illustrated by the graph in Figure 6-2*. A 0.5ms pulse can be translated to a duty cycle of 2.5%, a 1.5ms pulse translates to a duty cycle of 7.5%, and so on. The 20ms pulse is a given using the PWM pin, so the duty cycle easily translates to a pulse width.*

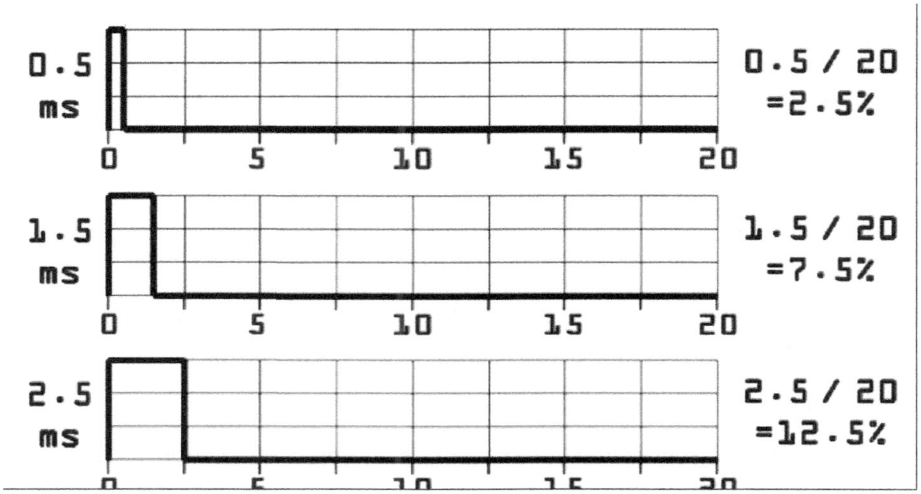

Figure 6-2. *Duty cycles*

To use PWM with the *RPi.GPIO* library, you need to connect the servo's signal wire (usually the white or yellow one) to one of the Pi's GPIO pins, and then use the `GPIO.PWM()` function. For example:

```
import RPi.GPIO as GPIO
import time

GPIO.setmode (GPIO.BOARD)

# This sets up GPIO pin 11 as our output pin
GPIO.setup (11, GPIO.OUT)

# This sets pin 11's frequency to 50Hz
p = GPIO.PWM (11, 50)

# This sets the duty cycle to 7.5 (centered)
p.start (7.5)

# To sweep back and forth, we change duty cycles
while True:
    p.ChangeDutyCycle (7.5)
    time.sleep (1)
    p.ChangeDutyCycle (7.5)
    time.sleep (1)
    p.ChangeDutyCycle (2.5)
    time.sleep (1)
```

Don't forget to power your servo with a separate battery (like a 9V for testing purposes) and to tie the external power supply's ground to your Pi's ground (pin 6).

Running this script should make the servo sweep left and right, pausing for a second between each direction change. However, if you run it with a servo attached, you'll notice that the servo movement is extremely jerky and noncontinuous. It's because, despite our best efforts, PWM output on the Pi is not as easy to manage as it is on a simple microprocessor such as the Arduino. While functional, it's not pretty. There's *got* to be a better way to do it.

ServoBlaster

Enter ServoBlaster, a *very* helpful library written by Richard Hirst. It runs unnoticed in the background as a daemon, and allows you to smoothly control a servo either within a Python script or even directly from the command line. It's also easy to download and install; you don't even need to use Git (the download/version control tool) to get it.

To install, first navigate into the directory where you're going to be putting all of your rover's Python code. Then, in the terminal, type this:

```
wget https://raw.githubusercontent.com/richardghirst/PiBits/master/ServoBlaster/user/servod.c
```

followed by:

```
wget https://raw.githubusercontent.com/richardghirst/PiBits/master/ServoBlaster/user/Makefile
```

You now have the two files you need. To install ServoBlaster, simply type the following:

```
make servod
```

To start it (as I said, it runs in the background), type:

```
sudo ./servod
```

You should see a screen full of code detailing the program's settings, including a servo mapping (Figure 6-3).

This mapping can be a little confusing, so it bears some explanation. First of all, you've probably noticed that the Pi's GPIO numbers don't correspond at all to its physical pin numbers (physical pin 4 is GPIO pin 2, for instance). When you use the *RPi.GPIO* library, therefore, you must specify whether you'll be referring to pins in your program by their physical pin numbers (with `GPIO.setmode(GPIO.BOARD)`) or by their GPIO numbers (with `GPIO.setmode(GPIO.BCM)`). The ServoBlaster library, unfortunately, adds another layer of complexity to this pin-mapping mess. The introductory splash screen tries to illustrate it, but with limited success. Referring back to Figure 6-3, you'll see that the eight servo pins that ServoBlaster uses are mapped to eight specific physical pins on the Pi: 7, 11, 12, 13, 15, 16, 18, and 22. Those *physical* pins, in turn, correspond to GPIO pins 4, 17, 18, 27, 22, 23, 24, and 25, respectively. ServoBlaster refers to those pins (servos) as servos 0, 1, 2, 3, 4, 5, 6, and 7, respectively. So if you want to control servo 0, you'll connect the servo's signal wire to the Pi's pin 7, also known as GPIO pin 4. If you want to control servo 5, you'll connect the servo to the Pi's pin 16, also called GPIO pin 23. And so on. Clear as mud? Don't worry, it will make more sense as you get familiar with the library. In the interest of keeping things simple, I use `GPIO.BOARD` mappings in this book, which has the added benefit of ensuring that the code here will run on whatever version of the Pi you happen to be using. If you're not using `GPIO.BCM` mappings, you can safely ignore the GPIO pin numbers I refer to.

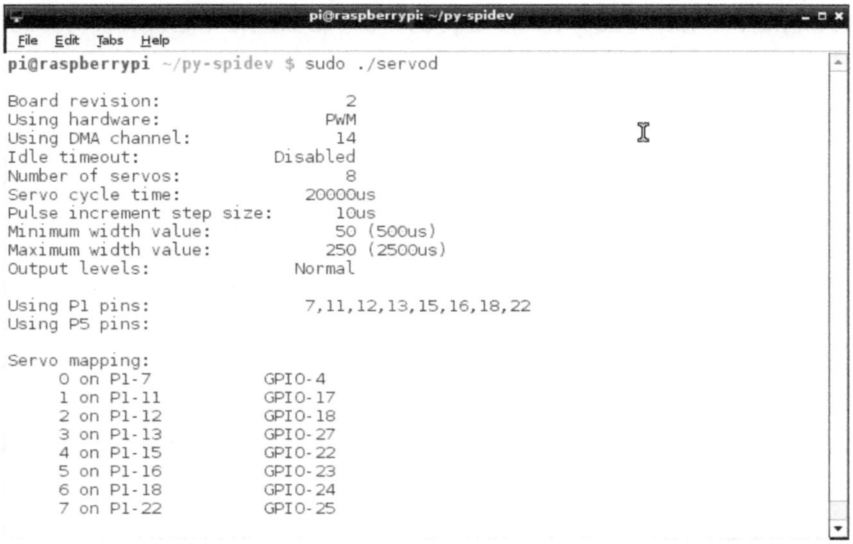

Figure 6-3. *ServoBlaster intro screen*

To test the library after you've installed it, connect a servo to your Pi's GPIO pin 18 (pin 12). Keep it powered with your 9V battery, and in your terminal, type:

```
echo 2=150 > /dev/servoblaster
```

Your servo should immediately do something interesting. The command follows the syntax `echo (servo=value) > /dev/servoblaster`. You're basically writing (`echo`) a value (`servo=value`) into the */dev/servoblaster* file. The library maps a servo number to a pin; if you want to refer to servo 0 in your code, it'll be the one plugged into pin 7 (GPIO pin 4) on the Pi. In order to use this syntax in our Python code, we'll simply use the *subprocess* library:

```
import subprocess
subprocess.call("echo 2=150 > /dev/servoblaster", shell=True)
```

You'll want to play around a bit with values you send to the servo, perhaps by writing a simple interactive Python script—an exercise I leave to you. Different servos respond differently; I've noticed discrepancies even between two servos of the same model and manufacture. Be aware, also, that only certain GPIO pins will work with the library. That is, the ServoBlaster library can control up to eight servos at a time, but only on the eight particular pins I mentioned previously. You'll need to investigate what values to send to the servo to make it move the way you need it to: depending on how you install your robotic arm, you'll need to know how far up to lift it and what value it takes to get it there.

Many thanks to Richard Hirst for his code library, written and freely given to the Pi community for projects like ours. If you need more information regarding the library and how to use it, check out his README file (*http://bit.ly/1tslnpi*).

7 Bot Construction

Now that we've gone over the history of Linux, and what the Pi is and what it can do, and how to set up the wireless, and (of course) the parts required, it's time to get down to some robot building. Remember, Rome was not built in a day, and neither will your rover. Take your time with the build, and plan out what you do before you do it. You know the saying: measure twice, cut once.

This is especially applicable if you are using the rover in this book as a suggestion, and not a step-by-step instruction list. With each modification you make (and you are encouraged to make them), you'll be faced with design decisions not mentioned here. I hope to prevent you from making some avoidable mistakes, but obviously I can't anticipate every possible design.

The Body

The most obvious place to start the build is with the body. Starting with the 4′ × 4′ piece of plywood, cut a piece large enough to become the base of the rover. Sketch it out or plan it on paper before you decide on size; it's going to have to fit two motors, at least two batteries (one for the Pi and at least one for the motors), a motor controller, at least one breadboard, the Pi itself, and several sensors. I made mine vaguely coffin-shaped, about 28″ × 15″ (Figure 7-1).

The shape is immaterial; if you want to go with a rectangular base to keep it simple, that's a great idea too. I was just feeling particularly motivated when I laid out the coffin design.

After you've designed and cut the base shape from the plywood, you'll next need to measure and cut the 1 × 4s to go around the edges. These are going to serve as the sides of the body. After you've cut them to fit, attach them to the base with screws or wood glue (Figure 7-2).

If you're going to attempt to make the body smooth (for aesthetics), it might behoove you to experiment with dovetails and 45-degree corner cuts to make the sides fit the base as snugly as possible. On the other hand, if you're not a carpenter, or are going for the "backwoods home-brewed robot" look, don't worry too much about how the sides fit. For that matter, you could conceivably build most of the rover on a flat base with no sides—just some upright portions through which to slide the axles.

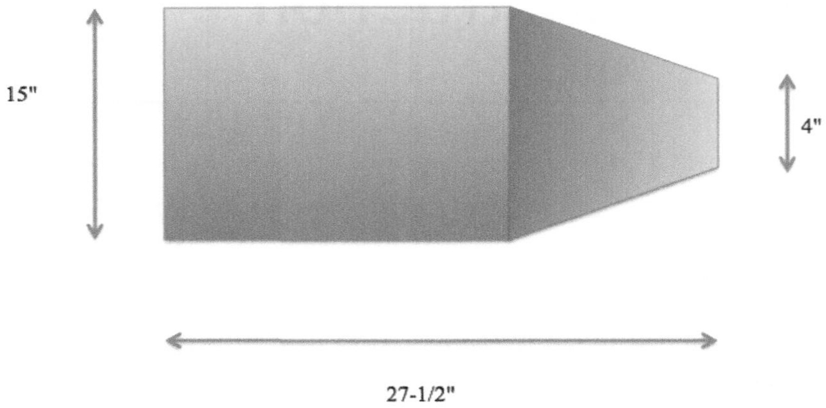

Figure 7-1. *Body shape of the rover*

Figure 7-2. *Attaching the sides*

When the sides are firmly attached, sand them as smooth as you can. If you have a lot of gaps and cracks, and if you're going for the smooth look, get some fiberglass paste (sold at auto parts stores for filling dents in auto bodies), and apply it to the cracks (Figure 7-3). Follow the instructions on the package to apply it, and after it sets and dries, sand it as smooth as you can. I normally do this sanding by hand, rather than with an electric sander, because it's much too easy to press too hard with an electric sander and remove more material than you planned. If you get a sanding block, not only will your surfaces be flat, but the block will help save your hands from fatigue. Start with a pretty rough grade to shave down the lumps, and then use finer and finer grade to get the surface smooth.

Figure 7-3. *Applying filler to the gaps*

Once it's as smooth as you can get it, quickly go over it with a damp sponge or cloth to wipe up the dust, as the dust will prevent paint from sticking properly to the surface. Finish up by painting the body with a primer *and* a final coat of paint. Feel free to decorate as you see fit (Figure 7-4).

Figure 7-4. *Finished body*

The next step in the construction of the body is drilling holes to hold your wheel axles. With the body and motor design I outline here, you'll want to keep the holes for the front axle as close to the actual size of the axle as possible to make it a tight fit. The rear axle holes, meanwhile, should be a bit larger to allow the axles to move as the rover moves (Figure 7-5)—sort of like shock absorbers. The rear axles will end up being made out of short pieces of flexible PVC pipe, so giving them some room to bend will allow the wheels to have more "give" when traveling over uneven ground. The axles will be rubbing against the wood, but the rover won't be going fast enough for the friction generated to affect its travel. If you prefer, however, you can find bearings that fit the axles and use them in the rear holes as well as on the front axle.

Figure 7-5. *Hole for rear axle*

Finally, if you plan to use a plexiglass cover for your rover, now is a good time to attach it. Using a plexiglass cutter (Figure 7-6) and a straight edge, cut it to fit the shape of the rover's body. You'll need to cut a bit out later to accommodate the robotic arm, but make sure it fits the body shape.

Figure 7-6. *Plexiglass cutter*

Don't attach the plexiglass to the body yet; it's much easier to cut while it's still unattached.

The Motors

The motors can now be prepared and then mounted to your robot's body by using brackets. The bracket must be thin enough to be attached with the motor's original screws, yet strong enough to hold the motor's weight. Spend some time shopping at the hardware store until you find a suitable bracket; I found these in the lumber section, as they're used to connect 2 x 4s when framing a house.

First, unscrew the faceplate from the motor and trace the hole positions onto the mounting bracket (Figure 7-7).

Figure 7-7. *Tracing the faceplate*

Now drill out the holes you traced onto the bracket. Make sure they're large enough to slip the faceplate screws through, and then remount the faceplate to the motor with the bracket attached. When you're finished, the faceplate screws should be not only holding the motor together, but also keeping the bracket tightly attached to the motor. This part of the project should end up looking something like Figure 7-8.

Figure 7-8. *Motors attached to mounting brackets*

The idea here is to securely fasten the motors to your rover's body, and the exact technique you use will depend on the motors and brackets you're using. If you're able to replace the faceplate screws with longer ones, you may be able to simply mount the motors to the side of the rover itself, eliminating the need for brackets. However, the motors I used have unique, tapered screws that can't be simply replaced with longer ones, so I had to find a thin, sturdy bracket option.

Finally, attach the brackets (with motor attached) to the rover base, so that the motor shafts extend through the holes in the side (Figure 7-9). Don't worry if they don't extend very far; we'll extend them a bit when we attach the wheels.

Figure 7-9. *Mounted motors*

The Wheels

Again, depending on the drive design you choose, this portion of the build may differ significantly from mine. However, if you're also going for the direct-drive approach, you should go for something similar to this. Each of the two rear wheels is attached to a motor, which is controlled via software running on the Pi. Meanwhile, the two front wheels are attached to the front axle with bearings, meaning each wheel turns independently of the axle. This means that in order to turn the rover, you need to turn only one rear wheel or the other (or both in opposing directions), and the front wheel on each side will follow suit. It also means that the front axle must *not* turn with the wheels, or the rover will simply move forward, rather than turning. Keep the front axle snug in its hole, don't be afraid to affix it permanently with glue or epoxy, and make sure the front wheels turn without problems.

Another possible wheel design (which I experimented with in theory as well) is a chain or belt drive. In this configuration, you need only one drive motor, which is then attached to the axle. A chain drive requires a sprocket or gear attached to the axle, while a belt drive will require some method of keeping the belt centered on the axle without slipping.

In my opinion, there are two disadvantages to the chain- or belt-drive model. First, keeping the belt or chain attached to the rear axle is a challenge; a sprocket for a chain would need to be solidly attached to the axle, and a belt would tend to slip on the axle unless you could devise a way of keeping it from doing so.

The second disadvantage is that if the rear wheels turn together, as they do in this configuration, then you must be able to steer the front wheels, either by twisting the entire axle or by turning the wheels themselves, the way the wheels turn on a full-sized automobile. Both of these present significant design challenges that you may or may not want to attack.

I prefer the direct-drive design, for the main reason that it lends itself well to being converted to a track design, like a tank. Connecting a track between the front and rear wheel on each side would make the rover even easier to steer and perhaps make it more of an all-terrain vehicle.

The Rear Wheels

Each rear wheel is solidly attached to a length of small plastic pipe, with a diameter just large enough to slip over the motor's shaft. To accomplish that, a strong epoxy, designed to attach plastic to plastic and plastic to metal, comes in handy. First, if necessary, use a round file to enlarge the hole in the middle of the wheel just enough to accommodate the pipe (Figure 7-10).

Figure 7-10. *Hole for axle shaft*

Before you epoxy the axle, you'll want to create a sort of hubcap to help solidly attach the wheel. To do this, use a large fender washer with a notch cut to fit the wheel hub (Figure 7-11). The inner hole of the washer does *not* need to fit the plastic pipe, as you'll be gluing it to the end of that pipe.

Figure 7-11. *Original and modified hubcap*

To connect the wheel, slide the pipe through the hole in the middle and use epoxy to attach it (Figure 7-12).

Figure 7-12. *Axle pipe attached to wheel*

When the epoxy has set according to the instructions, flip the wheel over and glue the hubcap to both the wheel hub and the end of the axle (Figure 7-13). The notch not only gives more surface area for attachment, but also ensures that the wheel will turn when the axle shaft does.

Figure 7-13. *Attached hubcap*

The last step in the process of building the drive mechanism is to attach the axle to the motor shaft. There are two ways of doing this: chemically, with epoxy, or mechanically, with a cotter pin. I chose to use epoxy; if you'd rather use a cotter pin, make sure you have a sharp drill bit, as the motor shaft is likely to be hardened steel. Slip the pipe over the motor shaft and drill a hole through both. Then slide a cotter pin through the hole and bend the ends back. This method has the advantage that you can remove the axle shafts at any time.

If, on the other hand, you prefer to go the more permanent route, break out your trusty tube of epoxy again. Mix up a large batch and slather it generously over the motor shaft. When it's fully coated, slide the pipe over the shaft and let the glue set according to the instructions. When you finish, you should have something like you see in Figure 7-14.

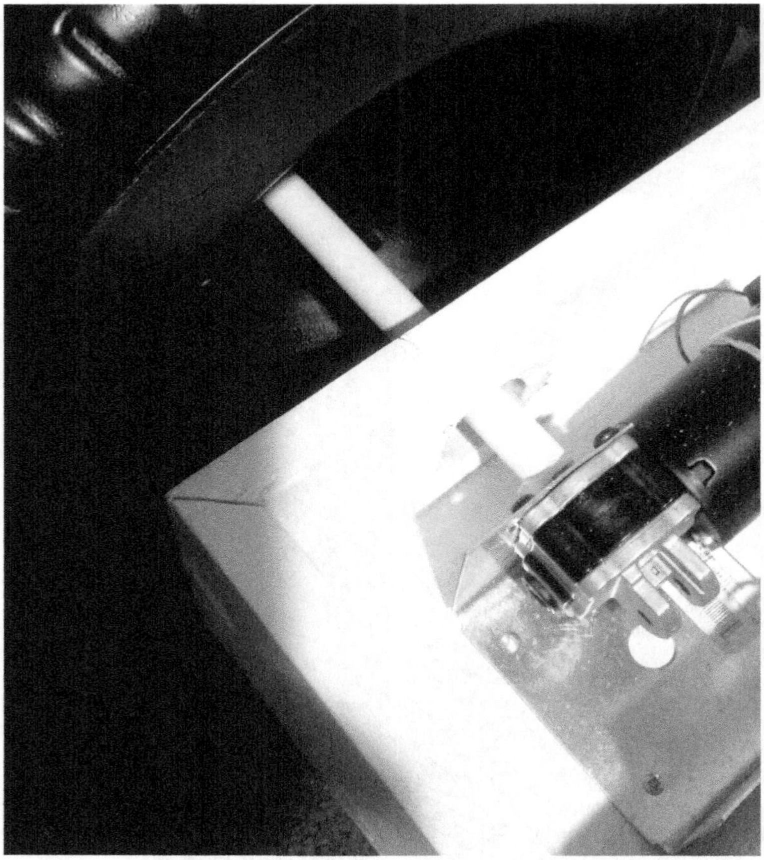

Figure 7-14. *Attached rear axle*

However you go about it, the end result should be an axle/wheel combination that turns when the motor shaft turns.

The Front Wheels

Luckily, the front wheel assembly is much easier, though again it uses epoxy. In order for the steering design to work, each front wheel must be able to turn without affecting the other front wheel. To accomplish this, they must be attached to the front axle with a bearing, and the axle itself must be firmly attached to the body of the rover. If you like, a viable alternative would be to use casters for the front wheels, as they don't require an axle and can turn independently of each other. If you decide to go with that approach, make sure you can mount them solidly to your rover's body and make sure that any off-road driving you intend to do won't gum up the bearings in the casters. I prefer to stay with four matching wheels, which requires freely rotating wheels on an immobile front axle.

Once again, mix up a batch of epoxy and, working quickly, slather it over the inside of the front wheel where the axle would sit. Before the glue sets, affix the metal bearing to the wheel; the end result should be a bearing that is firmly attached to the wheel and can be slipped onto the front axle. It is also possible to simply make sure the wheel turns on the axle freely and keep it in place with a few strategically placed cotter pins, but the disadvantage is that the wheel may wobble quite a bit, contributing to the rover's instability. Attaching the wheel to a bearing, which is then attached to the axle, ensures the wheels will roll smoothly and without wobbling.

When the epoxy holding the bearing to the wheel is firmly set, slide the axle through the holes you drilled for it in the body of the rover, and then slide the bearing over the axle and attach it firmly (Figure 7-15).

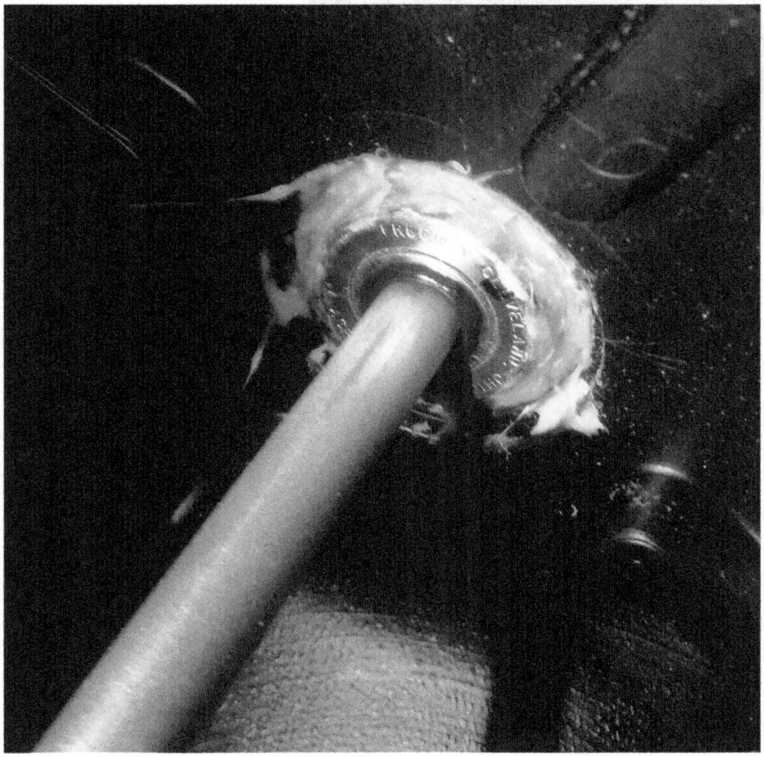

Figure 7-15. *Bearing and axle attachment*

I used cold-weld for this; it's similar to epoxy, but specifically made for bonding metal to metal (aluminum to steel, in our case—the axle to the bearing). The final product should look something like Figure 7-16.

Remember to slide the axle through the rover body before *you attach the bearings; otherwise, you'll have an axle that can't be attached to the body!*

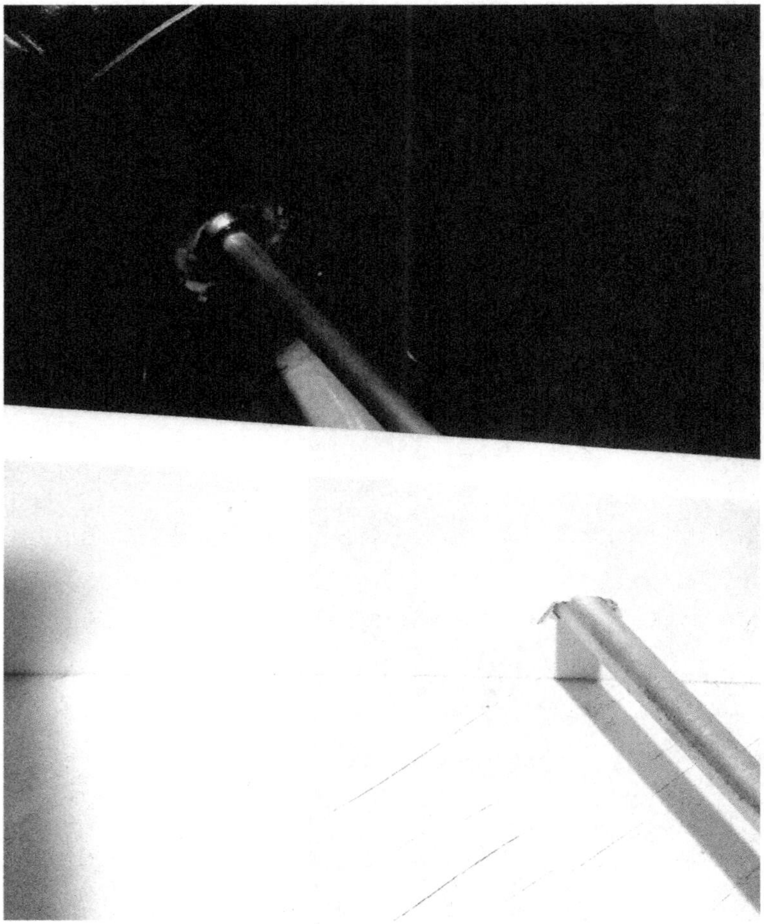

Figure 7-16. *Front wheel and axle*

The Robotic Arm

Perhaps the neatest thing about this rover is the robotic arm. You can mount a webcam or the Raspberry Pi camera board to the end of the arm, and either take still pictures with it, or perhaps even navigate using a live stream. The high-torque servo allows you to raise and lower the camera, giving you the ability to get a good look around your surroundings.

Start by determining where you're going to mount the arm on your rover's body. For stability's sake, you'll probably want to mount it toward the center or rear of the rover, on the midline. Then decide how long you want the arm, and cut the aluminum channel to that length. Remember that the longer the arm, the more torque will be required to raise it, as the servo will be mounted at the hinge point on the rover's body. The camera on the end will also contribute to the weight, as will the enclosure you design to hold it.

The high-torque servo will now need to be attached to a mounting bracket of some sort. I used a bracket similar to those I used to mount the rear motors. Cut a hole in the bracket that fits the servo snugly, and then slide the servo through and attach it to the bracket using the built-in mounting holes (Figures 7-17 and 7-18).

Figure 7-17. *Servo-shaped hole*

When the servo is firmly attached to the bracket, attach the length of aluminum channel to the servo's horn by using some sheet-metal screws (Figure 7-19).

Figure 7-18. *Servo attached to bracket*

Figure 7-19. *Arm attached to servo horn*

The whole assembly can then be mounted to the body of the rover (Figure 7-20).

Figure 7-20. *Mounted servo and arm*

After your servo is attached, you can concentrate on the *other* end of the arm, where the camera(s) are attached. You'll probably want some sort of enclosure, or at the very least a holder. To keep the weight down, I made the camera holder out of a block of craft Styrofoam. If you want to ensure that the camera is looking forward whether the arm is standing straight or laying forward on the body of the rover, attach the block of Styrofoam to the end of the arm, using a long threaded rod as a pivot. Carve out a section of foam for the end of the arm, and then mount the camera on the foam by using a bit of hot glue (Figure 7-21).

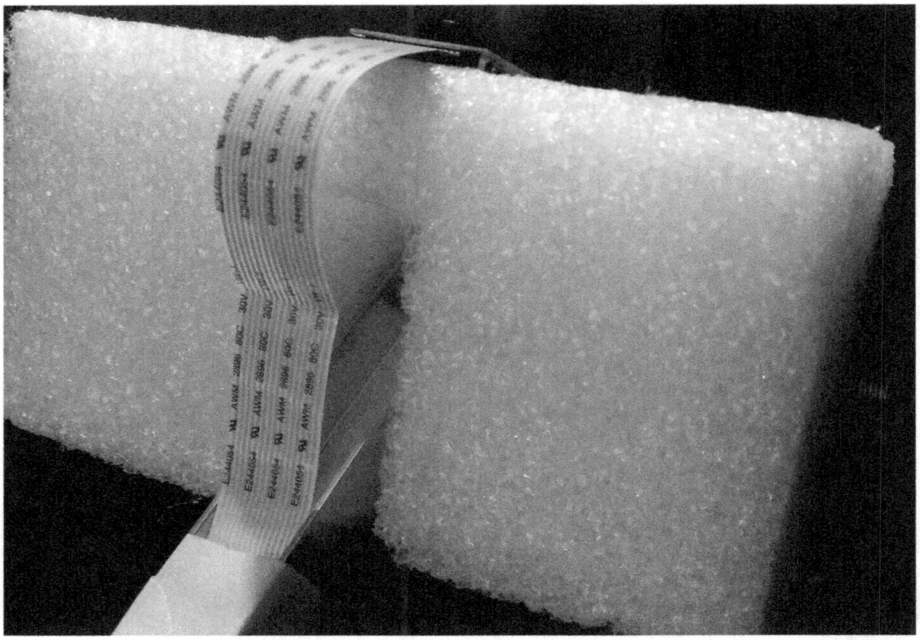

Figure 7-21. *Camera attachment to arm*

It doesn't take much glue to attach the Pi camera, because it weighs so little, and the glue won't damage the camera as long as you don't use too much. To ensure that the camera remains level, you may need to add some weight to the bottom of the enclosure. Lead fishing weights or even a few large screws work well for this.

When you're satisfied with the attachment and pivot, run your camera's cable down the inside of the aluminum channel (Figure 7-22), using some tape to hold it in place.

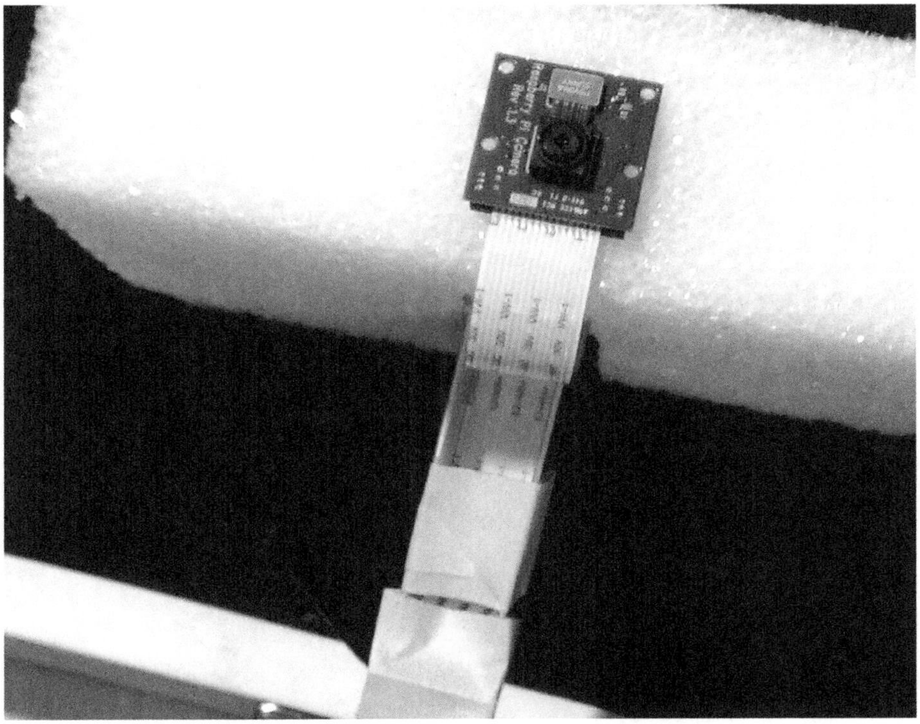

Figure 7-22. *Camera cable*

Finally, to hide the guts of the arm and its enclosed wiring, use a length of flexible plumbing hose, cut lengthwise to slide over the arm, with appropriately placed slits to accommodate the cable connectors where necessary (Figure 7-23).

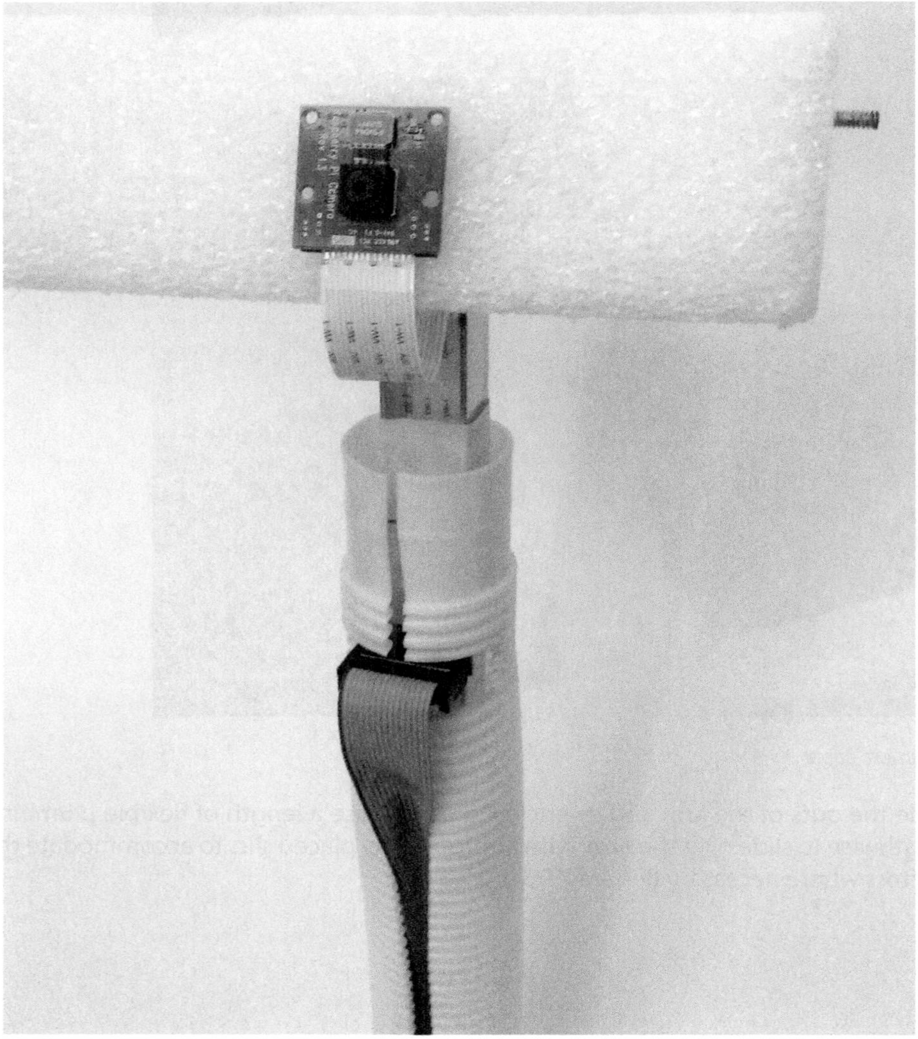

Figure 7-23. *Robotic arm cover*

Pi Power

In order to supply your Pi with its required 5V with a Li-Poly battery pack, you'll need to create some sort of connection between the battery pack and the mini USB cable used to plug into the Pi. You'll also need to incorporate a voltage regulator of some kind to bring the battery voltage down to a Pi-friendly level. In my experience, the easiest way to accomplish all of these things in a small package is to use one of the USB chargers that plug into a car's 12V socket (Figure 7-24).

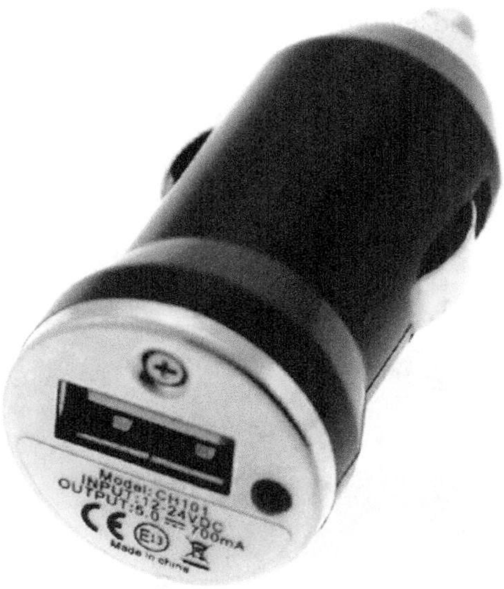

Figure 7-24. *USB charger*

The charger can be connected to the Pi with a short USB-to-USB mini cable. I use an extendable one. To attach the charger to your batteries, get an extension socket for the charger (Figure 7-25).

Figure 7-25. *Extension socket*

Cut off the end of the extension cord, and solder an XT60 female connector (or whatever type of connector matches your Li-Poly pack) to the end (Figure 7-26).

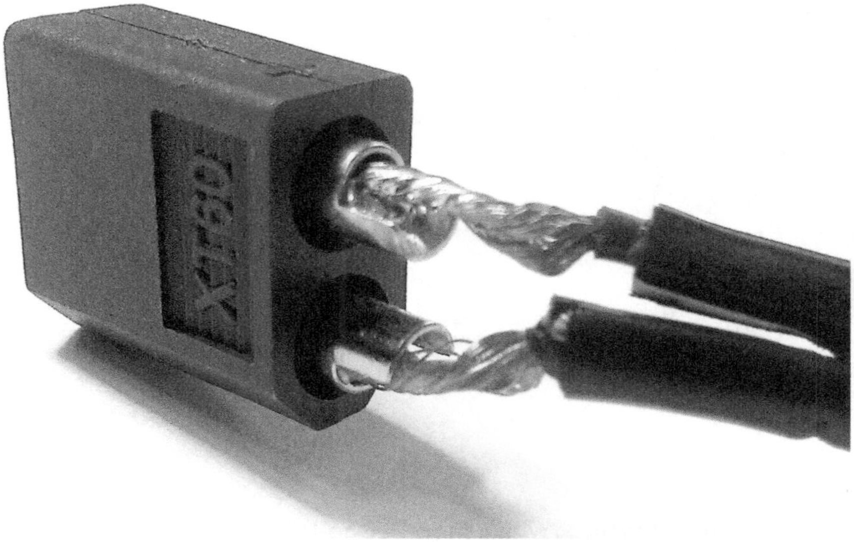

Figure 7-26. *Soldered connector*

When you're finished, you should have two self-contained power plugs that fit nicely together (Figure 7-27).

This allows you to plug and unplug battery packs as you need/go through them, and the voltage regulator keeps your Pi safe. All in all, it's a portable power solution that I've used in many projects.

Placing Everything

The last part of the construction of the rover is placing all of the parts inside, including the Pi, the L298H motor-controller board, the various sensors, and the batteries and breadboards used to power and connect everything. Obviously, this will depend heavily on your rover's design, plus where you decided to place everything. Try to keep the wiring neat and tidy, and color-code wires where possible—use red wires for positive signals, black for negative, white for signal wires, and so on. Zip ties can be handy when it comes to bundling wires and cables. Your rover should end up looking something like Figure 7-28.

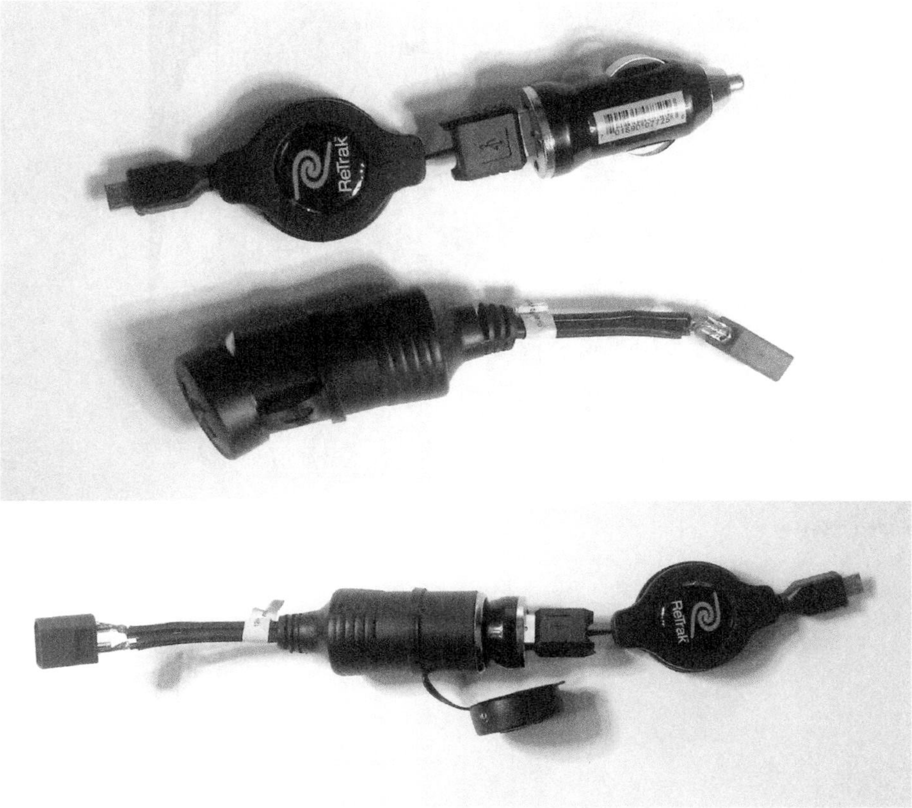

Figure 7-27. *Finished power cable*

Finally, consider adding a switch between the power for the motors and the L298H board. By using a simple toggle switch to disconnect the power when the rover isn't in use, you can prevent your large drive batteries from gradually leaking into the L298H board, prolonging their life. The Li-Poly packs used to power the Pi are easy to remove and recharge, but recharging the batteries I chose for the drive motors is a bit more involved; I use a car battery charger, and the less often I have to do it, the better.

Now that your rover is built, let's look at adding some sensors to it.

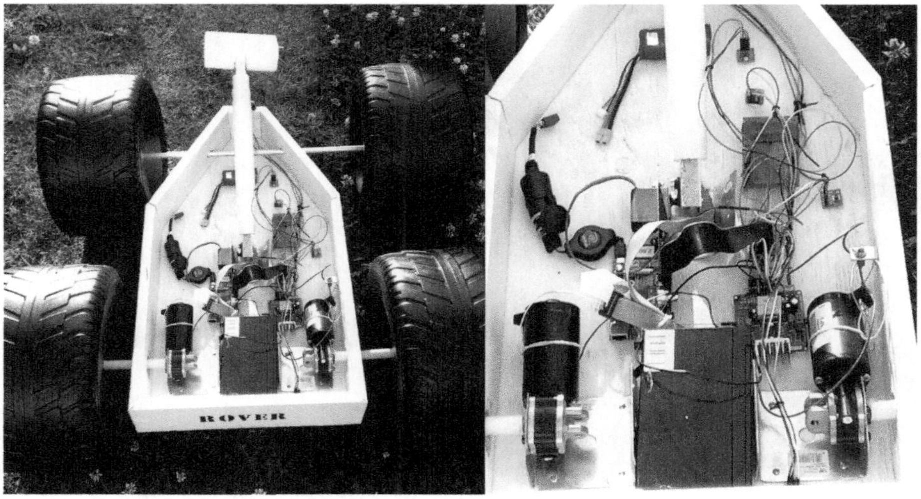

Figure 7-28. *Finished rover*

8 Bot Control

At this point in the build, after we hook up the motors and connect the motor controller to the Pi, we could remotely log into the rover and control it from a laptop. If you were to add a battery for the Pi and at least one for the motors, the rover would be technically ready to test outside. But don't do that yet. It might behoove us to connect everything *first*, debug the inevitable wire crossings and incorrect hookups, and *then* take it outside.

Connecting the Motors and Motor Controller

Setting the sensor connections and the robotic arm aside for a moment, let's go through the process of hooking up the motors so we can test the movement. You'll need to set aside six GPIO pins for the motor controller—three for each motor. I used pins 19, 21, 22, 23, 24, and 26, simply because they're six GPIO pins that are located close to each other, physically. I connected pins 19, 21, and 23 to the IN1, ENA, and IN2 inputs, respectively, and pins 22, 24, and 26 to the IN3, ENB, and IN4 inputs. As a reminder, each motor uses three pins: two input pins (labeled IN1 and IN2) and one enable pin (labeled ENA and ENB). Sending HIGH signals to different combinations of those three pins results in different motor behavior, as you'll see in Table 8-1. Arguably the most important pin out of the three is the enable pin; if an enable pin is set LOW, that motor doesn't turn, regardless of what signals are sent to the two input pins.

After you've connected the input and enable pins, connect the Pi's 5V output (pin 2) to the +5V pin on the controller, and the GND pin to the Pi's GND pin (pin 6). Finally, connect the two leads from the right motor to the board's two MOT1 connections, and the two leads from the left motor to the MOT2 connections (Figure 8-1).

Figure 8-1. *Connections made to the motor controller*

As I mentioned in the preceding chapter, I also installed a switch between my drive battery and the L298H board, so I don't need to worry about phantom power drain, or some random signal noise making the motors go nuts. When I'm ready to run the rover, I simply flip the switch and get power to the board.

To control the motors using the L298H board, you enable a motor by setting that motor's EN pin high. Then you can either spin the motor one way by sending a HIGH signal to one IN pin and a LOW signal to the other, or reverse the signals and make the motor turn the other way. This might be better illustrated by Table 8-1.

Table 8-1. Motor settings

ENA Value	ENA = 1	ENA = 1	ENA = 1	ENA = 0
IN1 Value	IN1 = 1	IN1 = 0	IN1 = 0	-
IN2 Value	IN2 = 0	IN2 = 1	IN2 = 0	-
Result	Motor spins CW	Motor spins CCW	Motor stops	Motor stops

When you read "Motor stops," you can read that as "Motor screeches to a halt." In other words, there is no coasting here; when a motor stops, it *stops*.

Because we're using the *RPi.GPIO* library, it's a simple matter for us to send pins HIGH or LOW by using these commands:

```
GPIO.setup (PinNumber, GPIO.OUT)
GPIO.output (PinNumber, True)
GPIO.output (PinNumber, False)
```

If we do this based on input from us (the user), we can control the motors with an interactive Python script.

Once you've connected the motors and power, it's a good idea to get the rear tires off the ground so you can test different commands without having your rover run away from you. Because the rover is not moving, this setup also has the advantage that nothing has to be portable; you can just plug in your Pi and use an AC adapter (if you have one of the right size) for your motors. Only *after* you fix the bugs in your code (oh yes, there will be bugs) do you need to plug in batteries and run around after the rover as it goes all sorts of places you never intended. For now, think of it as having your rover up on the rack in the garage. Remember, you'll be hooking the L298H board to your Pi's pins as follows:

- IN1 → pin 19
- ENA → pin 21
- IN2 → pin 23
- IN3 → pin 22
- ENB → pin 24
- IN4 → pin 26

With the drive wheels elevated, start a new Python script called *motortest.py*:

```
import RPi.GPIO as GPIO
import time

GPIO.setwarnings(False)
GPIO.setmode(GPIO.BOARD)

#19 = IN1
#21 = ENA
#23 = IN2
GPIO.setup(19, GPIO.OUT)
GPIO.setup(21, GPIO.OUT)
GPIO.setup(23, GPIO.OUT)

#22 = IN3
#24 = ENB
#26 = IN4
GPIO.setup(22, GPIO.OUT)
GPIO.setup(24, GPIO.OUT)
```

```
GPIO.setup(26, GPIO.OUT)

def rForward():
   "R motor forward"
   GPIO.output(21, 1)
   GPIO.output(19, 0)
   GPIO.output(23, 1)

def lForward():
   "L motor forward"
   GPIO.output(24, 1)
   GPIO.output(22, 0)
   GPIO.output(26, 1)

def rBackward():
   "R motor backward"
   GPIO.output(21, 1)
   GPIO.output(19, 1)
   GPIO.output(23, 0)

def LBackward():
   "L motor backward"
   GPIO.output(24, 1)
   GPIO.output(22, 1)
   GPIO.output(26, 0)

def allStop():
   GPIO.output(21, 0)
   GPIO.output(24, 0)

rForward()
lForward()
time.sleep(2)
allStop()
time.sleep(0.5)
rBackward()
lBackward()
time.sleep(2)
allStop()
```

You can experiment with other sequences, of course, but running this simple script with

```
sudo python motortest.py
```

should result in your wheels going forward for 2 seconds, pausing, and then going backward for 2 seconds. Play around with other values until you're familiar with how your motors respond and how you have them hooked up. Again, keep this script, as you'll be using it in your final code.

Controlling the Robotic Arm

The other bit of code you'll need in order to control your rover is the code necessary to raise and lower the robotic arm. You may remember it from Chapter 6, but let's revisit it anyway by writing an interactive script so you can see exactly what values have what effect on your arm.

To keep things simple, you should be in the same folder in which you installed the ServoBlaster library. (You don't *need* to be, but it keeps everything simple and organized.) Make sure your `servod` program is running with `sudo ./servod`, and connect your servo to pin 12 (GPIO 18) on the Pi. If you get confused, refer back to Chapter 6, or reread the startup splash screen for the `servod` command (Figure 6-3). Now start a new Python program:

```
nano servomap.py
```

Then try entering and running the following program:

```
from subprocess import call
import time

while True:
    position = raw_input("Enter servo value: ")
    call ("echo 2=" + position + " > /dev/servoblaster", shell=True)
    time.sleep(2)
```

When you run this script, you'll have the ability to send different servo values to the arm to see where it ends up. Ideally, you'll probably want the arm to be almost resting on the front of the rover when it's not deployed, and you'll want it almost straight up when it's active (Figures 8-2 and 8-3).

In fact, play with the arm configuration to find the value that allows the arm to rest on the body of the robot when not in use. Putting the arm at this rest position will help save your batteries, as otherwise the servo will continue to draw power to keep the arm in position. When you put the arm into the rest position and then send a value of `0` to the servo with

```
echo 2=0 > /dev/servoblaster
```

it will power off the servo, saving on power.

My arm's values ended up being `45` at rest, and about `100` when standing at attention. Obviously, your values might differ significantly depending on your servo, your arm, and how it's placed in your rover.

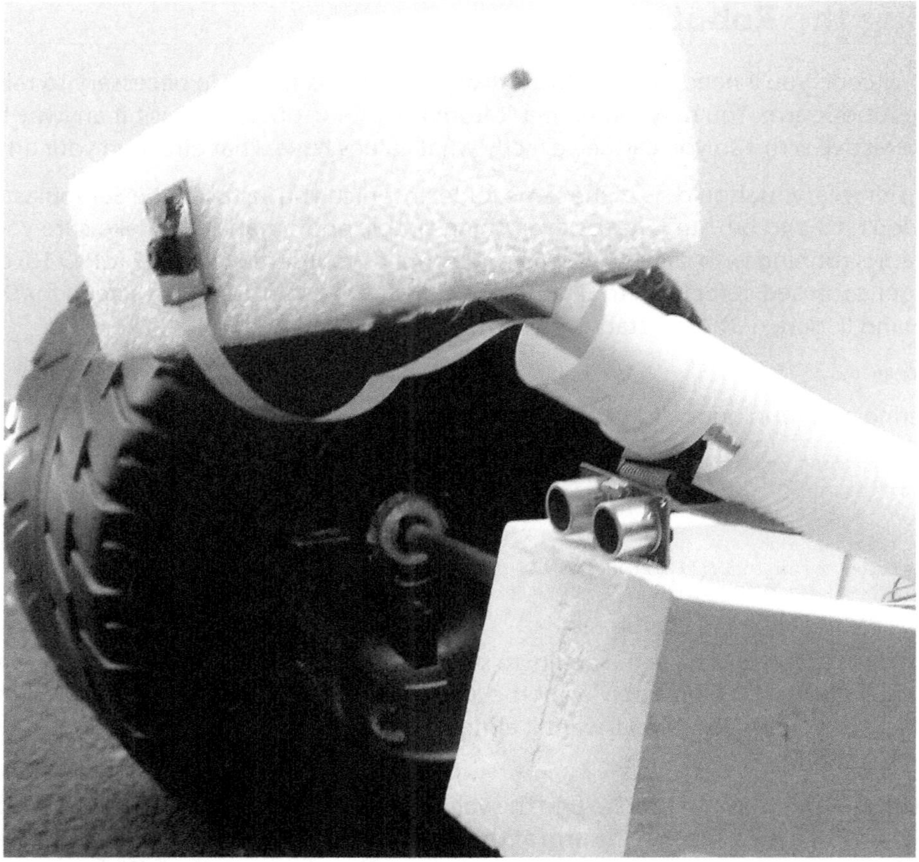

Figure 8-2. *Robotic arm at rest*

Now that you have your values, you need to write `raise arm` and `lower arm` functions to add to your final code. One thing you've probably noticed is the almost *violent* way the arm shoots from point A to point B. To keep things from breaking on the rover, and to maintain the illusion of robotic grace (as well as not to poke anyone's eye out), I wanted to slow things down a bit. To do that, I wrote a script that iterates slowly through the intervening values. To wit:

```
from subprocess import call
import time

def raise_arm():
   for i in range (45, 100):
       call ("echo 2=" + str(i) + " > /dev/servoblaster", shell=True)
       time.sleep(0.5)

def lower_arm():
   for i in reversed(range (45, 100)):
```

```
            call ("echo 2=" + str(i) + " > /dev/servoblaster", shell=True)
            time.sleep(0.5)
```

The script is pretty self-explanatory, but basically the `raise_arm()` function sends the values from 45 to 100 to the servo, one at a time, with a pause of half a second between movements. The `lower_arm()` function does the same thing in reverse. When you run this, you'll see your arm slowly raise and lower at a much more respectable pace—experiment with the `time.sleep()` values as you see fit.

Although the rover is a complex piece of machinery (it *is* a robot, after all), these two snippets of code are all the basics you need in order to control it: `move forward`, `move backward`, `turn left`, `turn right`, `raise arm`, and `lower arm`. In the next chapter, we'll go over using the GPS module to help you figure out where the rover is, and then in Chapter 10 we'll look at some of the sensor possibilities and how to write code for them.

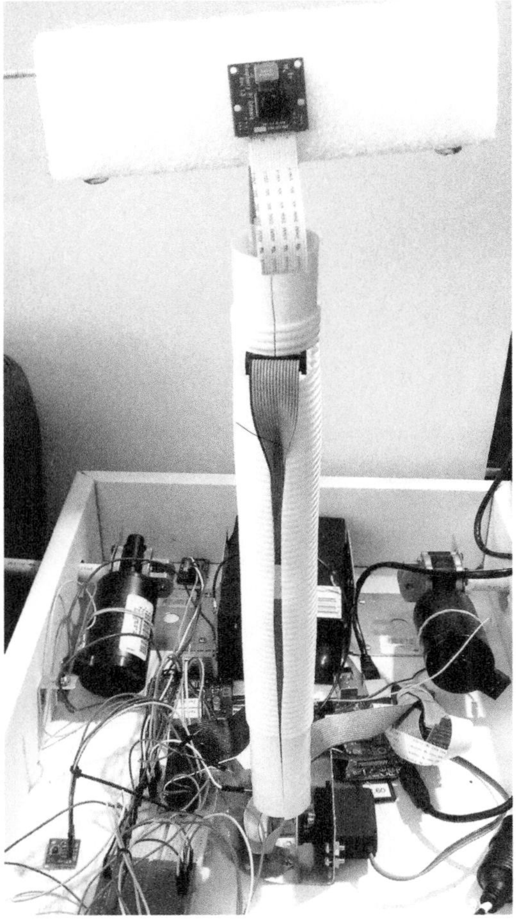

Figure 8-3. *Robotic arm at attention*

9 Bot Location

Once you've built a robot—particularly a mobile one, with wheels and perhaps a measure of autonomy—an important part of controlling it is knowing where it's gotten to while you weren't looking (see R2-D2's escape in *Star Wars Episode IV: A New Hope* for a classic example). While I don't think you'll need to install a restraining bolt on your rover, knowing where it is and where it's going, even to the extent of logging those locations to examine the route later in Google Maps or Google Earth, is definitely an exciting possibility. We'll be using a GPS unit, shown in Figure 9-1.

GPS stands for *Global Positioning System*, a satellite-based navigation system that can provide the location and time (in UTC format) for any place on Earth where there is a direct line of sight to four or more GPS satellites. It was developed in 1973 and started with 24 satellites; the system has since been expanded to 32 satellites, all owned and operated by the U.S. government. If a GPS receiver on the ground can lock onto a signal from these satellites, it can pinpoint a location on Earth with an accuracy approaching 1 meter (*http://1.usa.gov/1DsNFHe*), and a timing accuracy of around 100 nanoseconds (more than precise enough to determine whether your rover is safely across the street). It's often used in robotic systems, including military drones and Google's unmanned cars. GPS was originally established for the U.S. military, but in 2000 President Clinton directed the Pentagon to make the signals freely available and unencrypted. They now can be accessed by relatively inexpensive devices, which, in our case, can then be connected to the Raspberry Pi.

A GPS unit might be a little more expensive compared to some of the other sensors on the rover, but it's small, portable, and doesn't draw much power. It's also easy to set up and use. It can be purchased with an optional 5m antenna, but my experience is that the antenna isn't necessary; I can pick up enough satellites to get a GPS fix inside my house without it.

Figure 9-1. *GPS unit*

> *Fun fact: the GPS system must take into account Einstein's general theory of relativity during normal operations. Because ground-based observers are deeper within the Earth's gravity well than the satellites are, the clocks onboard the satellites appear to gain 38 microseconds per day. This discrepancy means that the satellite would misrepresent your position by about 10 kilometers! Because such a drift would make GPS useless in only a few minutes, this gain is corrected for in the system's design.*

Preliminary Setup

In order for the Pi to communicate with the GPS receiver, you'll first need to install the necessary Python library and its dependencies, and enable the correct interface on the Pi. The library is called *gpsd*, and it is available in the standard Raspbian repository. First, to install what you need, enter the following commands, one at a time, into your terminal, and let the package manager do its thing:

```
sudo apt-get update
sudo apt-get upgrade
```

```
sudo apt-get install gpsd
sudo apt-get install gpsd-clients
sudo apt-get install python-gps
```

(You may already have one or more of these installed, depending on your distribution version.) The Python gpsd module is part of a much larger library of code designed to allow devices such as the Pi and other microcontrollers to monitor and communicate with attached GPS and Automatic Identification System (AIS) receivers. It has been ported into C, C++, Java, and Python, and allows you to "read" the National Marine Electronics Association (NMEA)–formatted data transmitted by most GPS receivers. That means that if you decide to use a different GPS unit, either now or later, as long as that unit transmits the NMEA-formatted sentences, the Pi should still be able to understand the data. I have even had some success with a few off-the-shelf units that connect to the Pi via USB; you simply need to point the gpsd daemon to the USB connection (a bit further on in this chapter) rather than the UART one.

When you have the necessary packages and libraries installed, you'll need to configure the Pi's UART interface. *UART* stands for *universal asynchronous receiver/transmitter*. In simple terms, it is a piece of computer hardware that communicates information over a serial port, using the old reliable RS-232 serial protocol. The Pi has such an interface built in, preset to use pins 8 and 10 (GPIO 14 and GPIO 15).

By default, the Pi's UART interface is set up to connect to and communicate over a terminal window, but that configuration does us no good for reading the GPS, so we need to reconfigure it. To do that, start by making a copy of the */boot/cmdline.txt* file for safekeeping:

```
sudo cp /boot/cmdline.txt /boot/cmdlinecopy.txt
```

The *cmdline.txt* file is used to pass arguments to the Linux kernel. When you've saved a copy, edit the original:

```
sudo nano /boot/cmdline.txt
```

Its format is simply a space-delineated list of arguments. We need to delete the arguments that deal with terminal communications. In its original form, the file reads

```
dwc_otg.lpm_enable=0 console=ttyAMA0,115200
kgdboc=ttyAMA0,115200 console=tty1
root=/dev/mmcblk0p6 rootfstype=ext4
elevator=deadline rootwait
```

Delete this portion:

```
console=ttyAMA0,115200 kgdboc=ttyAMA0,115200
```

The file should now read (all on one line, obviously):

```
dwc_otg.lpm_enable=0 console=tty
root=/dev/mmcblk0p6 rootfstype=ext4
elevator=deadline rootwait
```

Save it, and then open the Pi's *inittab* file with the following:

```
sudo nano /etc/inittab
```

The *inittab* file describes what processes are started at bootup and during normal operation. All we need to do here is comment out the last line, which tells the Pi to start a terminal connection. The unedited line looks like this:

```
T0:23:respawn:/sbin/getty
-L ttyAMA0 115200 vt100
```

Add a hashtag to the beginning so the line looks like this:

```
#T0:23:respawn:/sbin/getty
-L ttyAMA0 115200 vt100
```

Save the file. Now, reboot the Pi with:

```
sudo shutdown -r now
```

Communicating with the GPS Module

When the Pi is back up and running, it's time to connect the GPS module and see what we can see. Though there are a lot of pins on the GPS unit's header board, we're using only four of them: VIN, GND, Tx, and Rx. Connect the VIN to the Pi's 5.5V, the GND pin to the Pi's GND, Tx to the Pi's Rx pin (pin 10), and Rx to the Pi's Tx pin (pin 8). It's easy to remember the direction of the last two connections: just remember that the Pi must *transmit* (Tx) to the board's *receive* pin (Rx), and it must *receive* (Rx) from the board's *transmit* (Tx).

When the red LED on the GPS board starts to blink, you'll know you have power. Now you can test it by starting the `gpsd` program. To do this, in a terminal, type this:

```
sudo gpsd /dev/ttyAMA0 -F /var/run/gpsd.sock
```

The `gpsd` program is a daemon, meaning it's a service that runs in the background, so you won't notice anything happening when it's running. The preceding command starts the `gpsd` program, tells the daemon that the GPS device is connected to the `/dev/tty/AMA0` interface, and that it should port the output of the device to the socket defined by `/var/run/gpsd.sock`. If you are connecting a GPS unit via USB, this command would instead read as follows:

```
sudo gpsd /dev/ttyUSB0 -F /var/run/gpsd.sock
```

Finally, you can start the generic GPS viewer/client by typing this:

```
cgps -s
```

The `cgps` client simply takes the data received from the `gpsd` program and displays it in a window. It may take a moment for data to stream, but you should see a window like that in Figure 9-2.

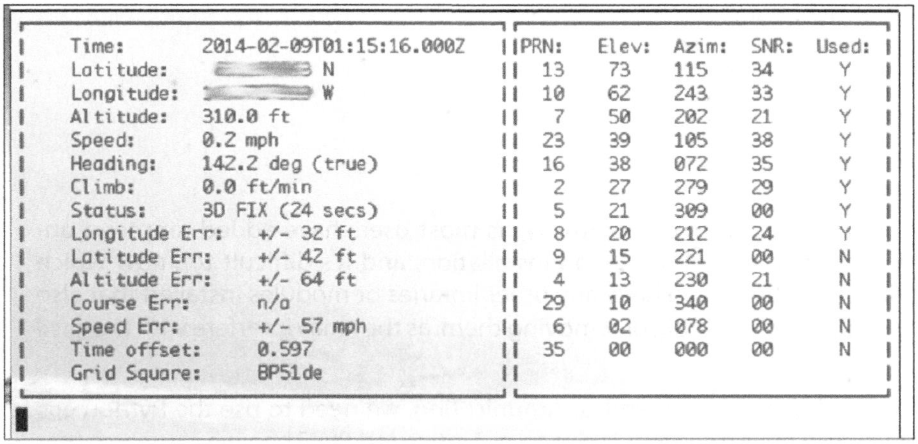

Figure 9-2. *cgps viewer window*

If you don't see any data, only zeros, it means the GPS can't find any satellites. The blinking LED will also slow down. One blink a second means the board is powered; as soon as it gets a fix on some satellites, the blink will slow to once every 15 seconds. Give it some time or a clearer view of the sky; as I said earlier, my experience is that the board is sensitive and you shouldn't have any problem obtaining a fix, even indoors.

If `cgps` always displays NO FIX and then aborts after a few seconds, you may need to restart the `gpsd` daemon. To do that, enter the following:

```
sudo killall gpsd
sudo gpsd /dev/ttyAMA0 -F /var/run/gpsd.sock
```

You may also want to start the `gpsd` program automatically, so it's running whenever you need it. To do that, open your *rc.local* file:

```
sudo nano /etc/rc.local
```

Right before the last `exit 0` line, add the previous `gpsd` command, so the last two lines of your file look like these:

```
gpsd /dev/ttyAMA0 -F /var/run/gpsd.sock
exit 0
```

Save the file and reboot your Pi, and the `gpsd` daemon should be running in the background.

If your GPS doesn't work immediately, don't get discouraged. I have found that sometimes uninstalling (and then reinstalling) the misbehaving modules does the trick, with the following:

```
sudo apt-get remove gpsd
sudo apt-get remove gpsd-clients
sudo apt-get remove python-gps
```

followed by:

```
sudo apt-get install gpsd
sudo apt-get install gpsd-clients
sudo apt-get install python-gps
```

Another thing you should *always* do is:

```
sudo apt-get update
sudo apt-get upgrade
```

Troubleshooting these modules can be tricky, as most users have added umpteen unknown packages to their default Raspbian installation, and it's difficult to know which modules play well together. If you have any other libraries or modules installed that also use the UART interface, try disabling or removing them, as they may interfere with the `gpsd` module.

Once we know the chip is working and communicating, we need to use the Python `gps` module to get useful values from the device to put into a log file. Then we can parse that log file and import it into Google Maps or Google Earth to see where our rover has been. The `cgps` client is handy, but not very useful for storing elements of GPS data such as latitude, longitude, time, and so on.

To use the GPS seamlessly, the easiest thing to do is to start reading its values in a separate thread. This is not a book on multithreaded programming in Python, and I don't want to go into too much detail as to how the code works, but a bit of information on threading might be helpful to you.

Threads enable a program to appear to accomplish seemingly endless numbers of things at once. Each task is spun off into a separate *thread* in the computer's memory, which runs independently of the other threads. The threads are not being executed *exactly* simultaneously, but the processor manages to switch back and forth between threads so rapidly that you think they're all happening at once. In Python, after importing the `threading` library, you can declare an object as a member of the `threading.Thread` class, with two included functions that initialize the thread and tell it what to do while it's running. Finally (if necessary), when the program ends, you can join the threads together and concatenate the results. In our GPS polling program, we just keep polling the GPS module and keeping track of our location—either printing it to the screen (for testing) or writing to a log file for import into another program later.

To test it, try the following script:

```
from gps import *
import time
import threading

f = open("locations.csv", "w")

gpsd = None

class GpsPoller(threading.Thread):
   def __init__(self):
```

```
            threading.Thread.__init__(self)
            global gpsd
            gpsd = gps(mode=WATCH_ENABLE)
            self.current_value = None
            self.running = True

        def run(self):
            global gpsd
            while gpsp.running:
                gpsd.next()

    if __name__ == '__main__':
        gpsp = GpsPoller()
        try:
            gpsp.start()
            while True:
                f.write(str(gpsd.fix.longitude)
                + "," + str(gpsd.fix.latitude)
                + "\n")
                time.sleep(30)
        except(KeyboardInterrupt, SystemExit):
            f.close()
            gpsp.running = False
            gpsp.join()
```

When you run this program, a *locations.csv* file will appear in the current directory, with a new line of updated location data every 30 seconds (obviously, if you're not moving, each line will be the same). We can play with the formatting and make it look however we need in order to import the data into another program, such as a mapping program. To that end, I'm using commas to delineate the data, as most online tools—should you decide to use them—import comma-separated values (CSV) files.

Using the GPS Data

So now you know you can communicate with the GPS module, and can access the data using the gps module's built-in functions (`.fix.latitude`, `.fix.longitude`, etc.) You'll need a Google account for the last part of this; once you have an account set up, open Google Maps (*http://maps.google.com*) and click My Places on the Settings Wheel at the bottom right of the screen (Figure 9-3).

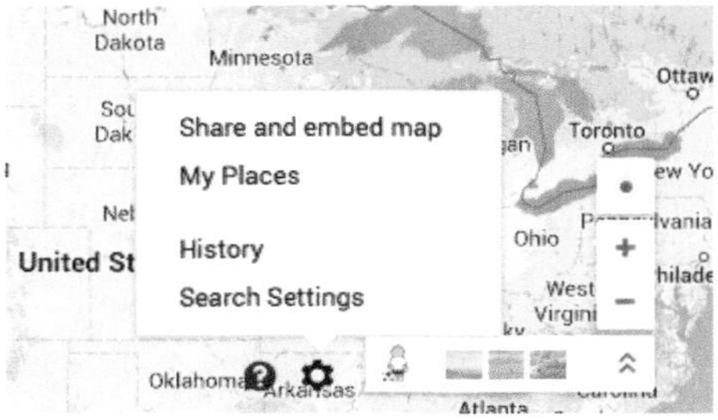

Figure 9-3. *My Places*

In the screen that opens, click the Create Map button, followed by the "Create a new map" button (Figure 9-4).

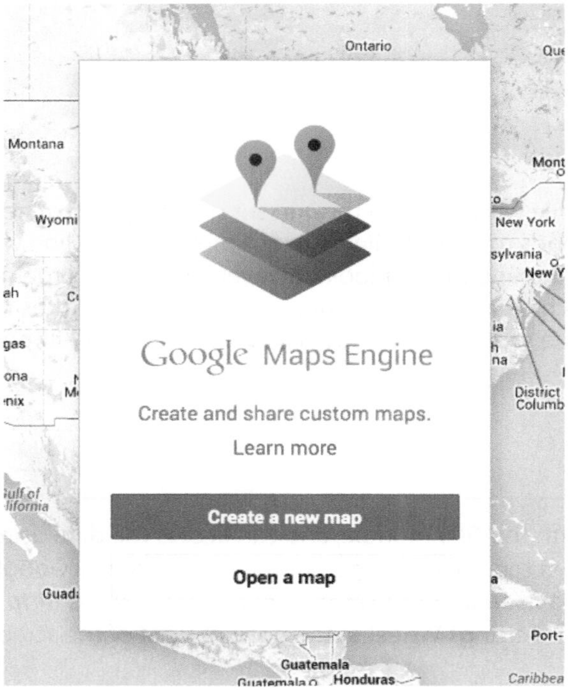

Figure 9-4. *Create a new map*

Follow the instructions to import/upload the *locations.csv* file you just created, and voilà! You now have a custom map created with the waypoints you recorded with your GPS. You can add columns as well to your logged CSV file; spend some time experimenting with the NMEA string that you receive from the GPS module and adding it to your data values.

If you'd like to see the path created in Google Earth (useful especially if you've utilized the gps module's .fix.altitude function), there's a bit more work involved unless you are a Google Earth Pro user. If you're not, you'll need to convert the data you've logged into a KML file. As it happens, you can do that relatively easily with a Python script.

A *KML file* is a special sort of eXtensible Markup Language (XML) file used by Google Earth to delineate landmarks, objects, and paths. It's formatted with opening and closing < > tags for different levels of information, such as <Document> and <path> and even <LineStyle>. By parsing our *locations* file line by line, we can create a descriptive KML file that can be recognized by and opened in Google Earth. Because we know how the final file needs to look, the parsing program can be written ahead of time, and you can just plug in your log file when your rover has returned home.

The first thing you'll probably have to do, however, is rewrite your GPS-logging script to save the locations, with spaces separating the latitude and longitude instead of commas (this has to do with line feeds, and parsing the file using Python, and will save you hours of debugging later). If you prefer, write to two files at once by adding a write line to the original script, writing to a *locations.log* file as well.

Your final *format_kml.py* file should look something like this:

```
import string

#open files for reading and writing
gps = open('locations.log', 'r')
#f = gps.readlines()
kml = open('locations.kml', 'w')

kml.write('<?xml version="1.0" encoding="UTF-8" ?>\n')
kml.write('<kml xmlns="http://www.opengis.net/kml/2.2">\n')
kml.write('<Document>\n')
kml.write('<name>Rover Path</name>\n')
kml.write('<description>Path taken by rover</description>\n')
kml.write('<Style id="yellowLineGreenPoly">\n')
kml.write('<LineStyle><color>7f00ffff</color><width>4</width></LineStyle>\n')
kml.write('<PolyStyle><color>7f00ff00</color></PolyStyle>\n')
kml.write('</Style>\n')
kml.write('<Placemark><name>Rover Path</name>\n')
kml.write('<styleUrl>#yellowLineGreenPoly</styleUrl>\n')
kml.write('<LineString>\n')
kml.write('<extrude>1</extrude><tesselate>1</tesselate>\n')
kml.write('<altitudeMode>relative</altitudeMode>\n')
kml.write('<coordinates>\n')
for line in gps:
        coordinate = string.split(line)
        print coordinate
        longitude=coordinate[0]
        latitude=coordinate[1]
        kml.write(longitude + "," + latitude + "\n")
kml.write('</coordinates>\n')
kml.write('</LineString>\n')
```

```
        kml.write('</Placemark>\n')
        kml.write('</Document>\n')
        kml.write('</kml>\n')
        gps.close()
        kml.close()
```

The first part of this code simply opens the necessary files and then writes all of the necessary KML formatting to *locations.kml* to set up the file. Then the small five-line loop reads through the *locations.log* file and places the data into the KML file. Then the script finishes up the KML formatting and cleans up after itself by gracefully closing the input and output files.

Your *locations.kml* file can now be viewed in Google Earth from any computer that has Google Earth installed. Right-click the file and from the Open With dialog box, choose Google Earth (Figure 9-5).

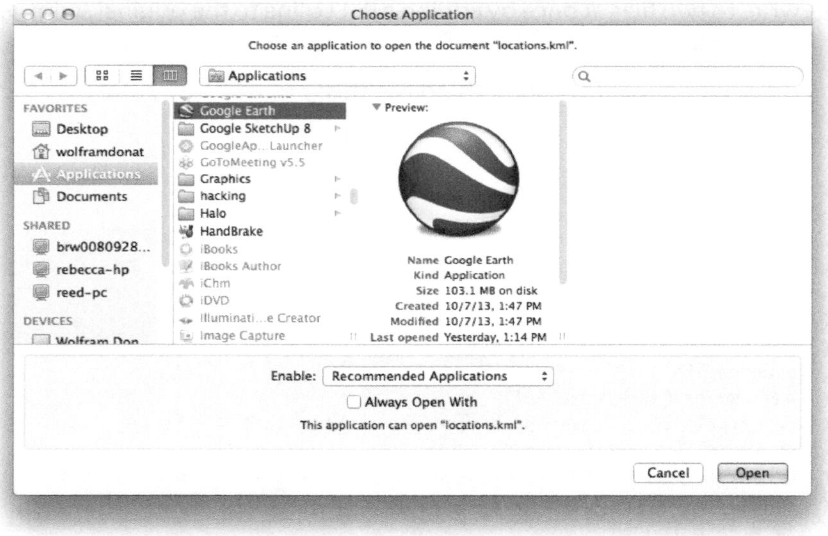

Figure 9-5. *Opening KML files on a Mac*

You should get a nice yellow path showing where your rover has been (Figure 9-6).

Figure 9-6. *Example rover path*

Of course, you can do tons of things with these location files, but one of the most interesting is to use them as waypoints in a preprogrammed route for the rover to follow. Plug in a sequence of points, and using the GPS module, your rover should be able to navigate to each of those locations and execute a preprogrammed sequence of events without any assistance from you. That, of course, is assuming that the terrain between point A and point B is navigable by the rover. Traveling between Flagstaff, AZ, and Cedar City, UT, for example, doesn't take into account the rather large canyon that lies between those two cities. (If your rover can successfully navigate the Grand Canyon autonomously, I'd be *very* interested in seeing it. So would NASA.)

This is just an introduction to the things you can do with your GPS module. In the next chapter, we'll take a look at adding sensors to the rover so you can do more than simply travel from point to point.

10 Sensors, Sensors, Sensors

When you're building a robot, some of its most important parts are its sensors. If the computer (the Pi, in our case) is the rover's brain, then the sensors you choose to pack into it are its eyes (a camera), ears (ultrasound), touch (reed switches and thermometers), equilibrium (accelerometers), and even senses that humans don't have, like magnetic field detection. The goal of most robots is not just to have sensors to receive input from the outside world, but to act on that input and do something. In the preceding chapter, we learned how to *remember* where our rover has been, and we touched on the possibility of autonomously navigating to those locations. But just navigating to them is kind of boring unless you can then use sensors to gather information from those points.

There are a lot of sensor possibilities you can use on your rover; I went over a few of them in Chapter 4, and if you spend a bit of time browsing some of the popular electronic sites such as Adafruit (*http://www.adafruit.com*) or Sparkfun (*http://www.sparkfun.com*) with a credit card and no spending limit, it's pretty much a sure bet that before long you'll have more sensors than you know what to do with. There are accelerometers and magnetometers and Hall effect sensors and reed switches and ultrasonic range finders and laser range finders, and the list goes on and on. In fact, you may be limited only by the size of your rover's body and by the number of GPIO pins available to you. (If you do find yourself running out of room, you may want to consider trying the Gertboard (*http://bit.ly/1t42VHm*) to expand your I/O potential.) In the meantime, two medium-sized breadboards should be plenty, depending on how well they fit in your rover's chassis.

In this chapter, I wanted to introduce some of the most common sensors and walk you through the process of setting them up and reading from them. Some sensors are quite basic—all you have to do is plug in electrical power and one or two signal wires, and you read the results. Others require libraries from third parties, like the SH15 temperature sensors. And some of them use the I2C protocol, a common way of communicating with external devices (sensors and other things) that are connected to the Pi. (For more on the I2C protocol, look to the end of this chapter.)

Obtaining and Using Sensor Code

An important thing to remember whenever you buy a sensor, especially from one of the aforementioned sites, is that often they have example or demonstration code available for download. Sometimes the code is only for the Arduino, as that's the original market for many of these sensors, but many times there is Python code for the Pi as well. If there isn't code, often a quick Google search will be fruitful—chances are you're not the first person to try to interface that sensor with the Pi. Remember: as a programmer, it makes no sense to reinvent the wheel, and there's no shame in reusing existing code.

And if you do end up having to write brand new code from scratch, consider sharing it with the Pi community. Someone else will surely get as much use out of it as you will.

Most of these sensors will come as fully assembled PCBs, but with the option of adding a row of soldered headers. I fully encourage you to solder the headers onto the board. It makes connecting and disconnecting each sensor via a breadboard a snap, and allows you to add to and subtract from your project easily. If you have more sensors than you can comfortably connect, having them all breadboarded will allow you to swap them out as needed.

If you're still new to soldering, fear not: soldering headers to a board is a good way to get used to the process, as the solder-phobic coating on the boards makes it pretty easy to keep the solder where it's supposed to be. Just remember: heat the joint, not the solder, and you should be fine.

The Art of Soldering

Possibly one of the most valuable skills to acquire if you're going to be working with electronics is the ability to solder well. At its core, it's really nothing more than melting a soft metal wire to securely connect two parts of a circuit together, but there is definitely an art to it, and just like any other skill, it comes more and more easily the more you practice. A few hours with a soldering iron can make a world of difference in your soldered joints.

There are four things to remember when soldering two pieces together:

- Prepare your surfaces: clean and rough surfaces make a better connection.
- Tin if necessary: melt a little solder onto a part to prepare it for connection.
- Connect the parts: make the connection mechanical if possible (twisting wires together, for example).
- Heat the parts: use the soldering iron to heat the *parts*, not the solder. When the joint is hot enough, the solder will flow easily, right where you want it.

There are all sorts of YouTube instructional videos on how to solder; Makezine.com has a particularly good resource-filled page at *http://makezine.com/2006/04/10/how-to-solder-resources*.

SHT15 Temperature Sensor

The Sensirion SHT15 is kind of a pricey sensor, retailing at about $30–$35, but it's also easy to install and use, which is why I recommend it. It doesn't use the I2C protocol (though it has pins labeled DATA and CLK, so it *looks* like it does); rather, you just plug the wires into your Pi, install the necessary library, and read the values it sends.

To use it, you'll first need to connect it. For our example, connect the Vcc pin to the Pi's 5V pin, and the GND pin to the Pi's GND. Then connect the CLK pin to pin 7, and the DATA pin to pin 11. You'll also need to install the *rpiSht1x* Python library by Luca Nobili. This is not a system-wide library, so navigate to within the directory where you'll be writing all of your rover code and download the library with

```
wget http://bit.ly/1i4z4Lh
```

When it's finished downloading, rename the file from the *bit.ly* name to what it should be with

```
mv 1i4z4Lh rpiSht1x-1.2.tar.gz
```

and then expand the result with

```
tar -xvzf rpiSht1x-1.2.tar.gz
```

Navigate into the resulting directory and install the library with

```
sudo python setup.py install
```

That should make the library accessible to you, so move up one level (back to your rover directory) and try the following script:

```
from sht1x.Sht1x import Sht1x as SHT1x
dataPin = 11
clkPin = 7
sht1x = SHT1x(dataPin, clkPin,
SHT1x.GPIO_BOARD)

temperature = sht1x.read_temperature_C()
humidity = sht1x.read_humidity()
dewPoint = sht1x.calculate_dew_point(temperature, humidity)

print ("Temperature: {} Humidity: {} Dew Point: {}".format(temperature,
humidity, dewPoint))
```

Save this with the filename *readsht15.py*, and run it with sudo:

```
sudo python readsht15.py
```

You should be greeted by a line of text describing your current conditions, something like

```
Temperature: 72.824 Humidity: 24.2825
Dew Point: 1.221063
```

This is the function you can use in your final rover code to access current temperature and humidity conditions.

Ultrasonic Sensor

The HC-SR04 ultrasonic range finder on the rover uses ultrasound to determine the distance between it and a reflective surface such as a wall, a tree, or a person. The sensor sends an ultrasonic pulse, listens for the echo, and then measures how long it took to receive that echo. The HC-SR04 is easy to configure on the Pi, and needs only two GPIO pins in order to work: one output pin and one input pin. The one caveat is that you should place a 1K resistor between the sensor and the Pi's input pin, as the HC-SR04 outputs 5V. This might damage the Pi's pin, as the Pi expects 3.3V as inputs, so a resistor brings the incoming voltage down to a Pi-safe level.

To power the HC-SR04, connect the sensor's Vcc and GND pins to the Pi's 5V and GND pins, respectively. Then choose two GPIO pins to be the *trigger* and *echo* pins. The trigger pin will be an output, and the echo pin will be an input. According to its datasheet, a quick ON pulse of 10 microseconds to the trigger pin will trigger the sensor to send eight ultrasonic 40KHz cycles and listen for the return on the echo pin. Luckily for us, Python's *time* library can be used to send microsecond-long pulses.

For the purposes of experimentation, let's hook up the sensor to pin 15 as the trigger and pin 13 as the echo. You can then use the following code to test the range finder:

```
import time
import RPi.GPIO as GPIO

GPIO.setmode(GPIO.BCM)
GPIO.setup(23, GPIO.OUT)
GPIO.setup(24, GPIO.IN)

def read_distance():
    GPIO.output(23, True)
    time.sleep(0.005)
    GPIO.output(23, False)

    while GPIO.input(24) == 0:
        signaloff = time.time()

    while GPIO.input(24) == 1:
        signalon = time.time()

    timepassed = signalon - signaloff
    distance = timepassed * 17000
    return distance

while True:
    print 'Distance: %f cm' %read_distance()
```

After importing the necessary libraries, we set up the GPIO pins to read and write. Then, in the `read_distance()` function, we send a 10-microsecond pulse to the sensor to activate the ultrasonic cycles.

After the pulse is sent, we listen for the echo on pin 27 until we get a 1 (meaning an echo has been sensed). We then mark how long it's been since the pulse was sent, multiply that by 17,000 (as directed by the README file) to convert it to centimeters, print it, and repeat with a `while` loop.

When you save and run this script (as `sudo`, remember, because you're accessing the GPIO pins), you should be able to wave your hand in front of the sensor and watch the values change with the distance to your hand. Keep this code handy, as the `read_distance` function will be used in our final code.

Photoresistor

A useful sensor to have in various situations, a *photoresistor* is a resistor whose resistance changes depending on the amount of ambient light (Figure 10-1).

Figure 10-1. *Photoresistor*

The photoresistor exhibits photoconductivity; in other words, the resistance *decreases* as the amount of incident light *increases*. Many of these resistors have a phenomenally wide range: nearly no resistance in full-light conditions, and up to several megaohms of resistance in near-dark. Photoresistors can tell you that your rover is working at night, for instance, or that it's in a tunnel, or that it's been eaten by a velociraptor.

The one caveat to using a photoresistor, however, is that you can't just plug it into your Pi and expect to read values from it. The photoresistor outputs analog values, and the Pi understands only digital ones. Thus, in order to convert its values into something the Pi can understand, you must hook it up to an analog-to-digital converter, or ADC chip.

The chip I use is the MCP3008, an eight-channel 10-bit converter IC available from several online retailers for around $5. Yes, you can easily get a more precise chip, with a 12- or a 16-bit converter, but unless you're measuring something that requires extreme precision, I don't think it's necessary. All we need are ambient light levels, and 10-bit precision seems to be plenty. It has 16 pins; pins 1–8 (on the left side) are the inputs, and pins 9–16 are voltage, ground, and data pins (Figure 10-2).

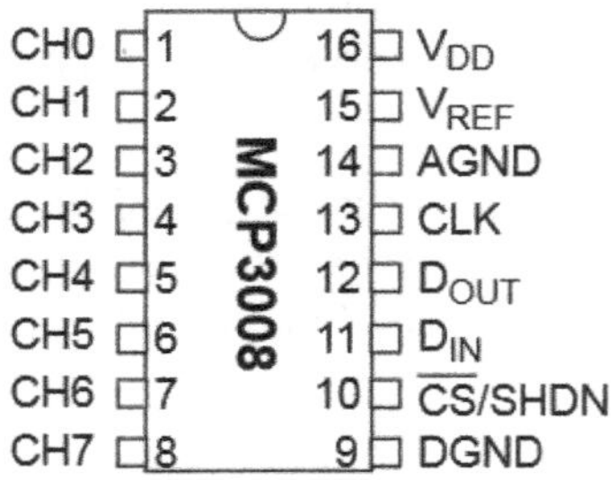

Figure 10-2. *MCP3008 pinout*

To read from the photoresistor, we'll use only one of the MCP3008's inputs, leaving seven inputs free for other analog devices we might wish to add later. The chip uses the SPI bus protocol, which is supported by the GPIO pins.

To use the chip and the photoresistor, you'll need to start by enabling the SPI hardware interface on the Pi. You may need to edit your *blacklist* file, if you have one. The *blacklist.conf* file is used by the Pi to prevent it from loading unnecessary and unused modules, and it exists only in earlier versions of Raspbian. Later versions don't have any modules blocked. See if */etc/modprobe.d/raspi-blacklist.conf* exists on your Pi; if it does, comment out the line

```
spi-bcm2708
```

and reboot. After rebooting, if you type **`lsmod`**, you should see `spi_bcm2708` included in the output—most likely toward the end.

You'll then need to install the Python SPI wrapper, with a library called *py-spidev*. First make sure you have the `python-dev` package:

```
sudo apt-get install python-dev
```

Then navigate to your rover's main directory, and download and install *py-spidev* with

```
wget https://raw.github.com/doceme/
py-spidev/master/setup.py
wget https://raw.github.com/doceme/
py-spidev/master/spidev_module.c
sudo python setup.py install
```

The library should be ready for use.

Now you can connect the chip to the Pi. Pins 9 and 14 on the MCP3008 should be connected to GND on the Pi; pins 15 and 16 should be connected to the Pi's 3.3V (you'll need to use one of the power rails on your breadboard). Then connect pin 13 to GPIO 11 on the Pi, pin 12 to GPIO 9, pin 11 to GPIO 10, and pin 10 to GPIO 8. Finally, for testing purposes, connect one leg of the photoresistor to pin 1, and connect the other leg to the ground rail.

When it's connected, all you need to do is open an SPI bus with a Python script and read the values from the resistor. The script should be something like this:

```
import time
import spidev

spi = spidev.Spidev()
spiopen(0, 0)

def readChannel(channel):
   adc = spi.xfer2([1,(8+channel)<<4,0])
   data = ((adc[1]&3) << 8) + adc[2]
   return data

while True:
   lightLevel = readChannel(0)
   print "Light level: " + str(readChannel(0))
   time.sleep(1)
```

Running this script (as `sudo`, because you're accessing the GPIO pins) should result in a running list of the values of the light hitting the resistor. The `readChannel()` function sends three bytes (`00000001`, `10000000`, and `00000000`) to the chip, which then responds with three different bytes. The `data` is extracted from the response and returned. You can test the system by waving your hand in front of the resistor to block out some light. If the resistance value changes, you know that everything is working properly. As with the other scripts, keep this one in your main rover folder for use in your final program.

Magnetic Sensor

The magnetic sensor, or Hall effect sensor, is a nifty little device that doesn't have a whole lot of uses outside of a certain small niche of applications. While I'm not certain it may be used on the rover, you may find yourself wanting to know if you're parked in the middle of a strong magnetic field (Figure 10-3).

Figure 10-3. *Hall effect magnetic sensor*

At its core, the Hall effect sensor is nothing more than a simple switch, so no real programming is required, nor are there any special libraries. To test it, connect the red (Vcc) pin to pin 2 on the Pi, connect the black pin to the Pi's GND pin (6), and connect the white signal wire to pin 11. Now just try the following code:

```
import time
import RPi.GPIO as GPIO
GPIO.setwarnings(false)
GPIO.setmode(GPIO.BOARD)
GPIO.setup(11, GPIO.IN, pull_up_down = GPIO.PUD_DOWN)
prev_input = 0
while True:
   input = GPIO.input(11)
   if ((not prev_input) and input):
       print "Field changed."
   prev_input = input
   time.sleep(0.05)
```

This script continually polls pin 11 to see if the input has changed (between `prev_input` to `input`). When you run it, your terminal should remain blank until you wave a magnet near the sensor. (You might have to experiment with different proximities and speeds; my experience is that the magnet has to come within a few inches of the sensor to register as a "pass"). Once you have it working, you can mount the magnet wherever you'd like to sense it, and poll the Hall effect sensor. When the magnetic field has changed, you know your robot is next to a magnet.

Pull-Ups and Pull-Downs

The fifth line in the magnetic sensor code bears some explanation, because it's an important concept that you'll come across often in your dealings with sensors. Whenever you set up a GPIO pin as an INPUT and connect a sensor to it, the pin becomes what's known as a *floating input* unless you do something special with it. This means that until a value is registered at that pin, the value of that pin could be anything at all: 0 volts, 3.3 volts, or anything in between. You can't predict (or act upon) such a floating input value. You need a way to set that floating input to a known value, like 0, until a true input value is read. In a circuit, this is normally done with a pull-up or pull-down resistor—a resistor that connects the pin to either GND or Vcc. This "pulls" the value of the pin to either LOW or HIGH, respectively, until a "real" value is read. Luckily, with the Pi's GPIO library, we have the ability to do that in code, with `pull_up_down = GPIO.PUD_DOWN`. This line means that if no value is read at pin 11, it will read 0. Likewise, `pull_up_down = GPIO.PUD_UP` pulls the pin to 1.

Reed Switch

Probably the easiest switch to program and connect is the *reed switch*, also called a *snap-action switch* (Figure 10-4).

The reed switch is often used by robots to determine the limits of some form of motion, from obstacle avoidance to grip control. It's a simple concept: the switch is normally open, permitting no current to pass. When an object presses on the switch, current is allowed through and the connected GPIO pin registers an INPUT.

Because you're using a physical switch, it's important that you're familiar with the concept of *debouncing*. Switches are often made of springy metal, and that can cause them to quickly "bounce" apart one or more times when they're first activated, before they finally close. An old analog circuit wouldn't register that "chatter," but a processor such as the Pi has no problem registering them. In the microseconds after contact is made, the Pi might be reading something like ON, OFF, OFF, ON, ON, OFF, ON, OFF, ON, OFF, ON, ON, OFF, ON, ON, OFF, ON, before settling down to a steady ON state. You and I know that the switch was triggered only once, but to the Pi, there was a festival of multiple switch triggers. This can cause problems when the program is running: did the robot run into an obstacle once, or dozens of times? The solution lies in software. By telling the Pi to wait until the chatter has quieted before reading a value, we debounce the switch.

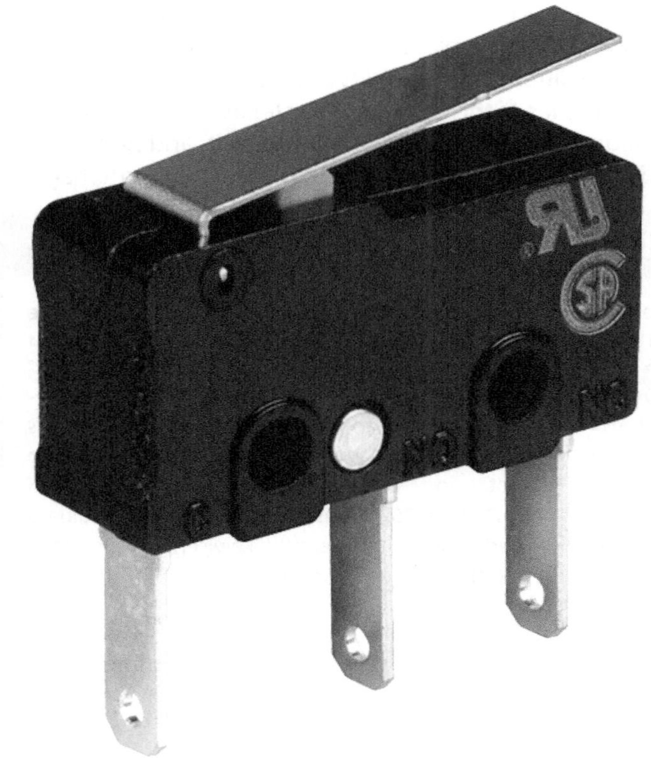

Figure 10-4. *Reed switch*

Early versions of the GPIO library made you do this yourself, in code, but later versions give it to you as a built-in capability. To demonstrate, I'll also show you the concept of an *interrupt*, where the Pi will stop whatever it's doing and alert you when something interesting happens at a pin.

Start by connecting the switch to your Pi. Wire one side of it to pin 11 (GPIO 17) and the other to the Pi's GND pin. If we pull the pin HIGH with a virtual pull-up resistor, then the input pin will register a switch close as being connected to ground. All that's left is to write some code to read it:

```
import time
import RPi.GPIO as GPIO
GPIO.setmode(GPIO.BOARD)

GPIO.setup(11, GPIO.IN, pull_up_down=GPIO.PUD_UP)

def switch_closed(channel):
   print "Switch closed"
```

```
# This next line is the interrupt
GPIO.add_event_detect(11, GPIO.FALLING, callback=switch_closed, bouncetime=300)


while True:
   time.sleep(10)
```

In the `switch_closed(channel)` function, we define what we want to happen when a switch close is detected. Then we use the `GPIO.add_event_detect()` function to look for a falling edge, with a bounce time of 300 milliseconds. (A *falling edge* is what happens when a signal that *was* high goes low; similarly, when a formerly *low* signal goes *high*, it's a *rising edge*.) When the main loop is started (all it does is wait, in this simple demo script), it waits for a falling edge to be detected on GPIO 17, at which point it calls the callback function.

The debounce and the interrupt are both handy tools for your robotic toolkit. Keep this code handy in your rover folder.

Motion Sensor

The motion sensor I use in my projects, the Parallax RB-Plx-75, works by sensing changes in the infrared "background" of its field of view (Figure 10-5).

Figure 10-5. *IR motion sensor*

Changes in the infrared signature cause the sensor to output a HIGH signal on its output pin. It's not so much a motion sensor or an IR sensor as it is a combination of the two. Like the magnetic sensor, it's really nothing more than a fancy switch, so connecting and programming it is quite simple.

It has three pins: Vcc, GND, and Output. Looking at the sensor from the point of view of Figure 10-5, the pins are OUT, (+), and (-). It can use any input voltage from 3V to 6V, so connect the GND pin to the Pi's GND, the Vcc to either 3.3V or 5V, and the OUT pin to pin 11 (GPIO 17) for this example. To test our code, let's connect an LED to signal if the sensor is tripped. Connect pin 13 (GPIO 27) to a resistor, and then connect the positive lead of an LED to the resistor's other leg. Finally, connect the LED's negative lead to GND on your Pi. You should now be ready to try the following script:

```
import RPi.GPIO as GPIO
import time

GPIO.setwarnings(False)
GPIO.setmode(GPIO.BOARD)

GPIO.setup(11, GPIO.IN, pull_up_down=GPIO.PUD_UP)
GPIO.setup(13, GPIO.OUT)

while True:
   if GPIO.input (11):
       GPIO.output (13, 1)
   else:
       GPIO.output (13, 0)
```

That's it! When you run the script (remember to use `sudo`, because you're accessing the GPIO pins), you should see the LED light when you move your hand around the sensor, and then go out again when there is no movement for a few seconds. Keep this code in your main rover folder.

---

*When working with LEDs, you should **always** use an inline resistor to limit the current passing through the LED. It's very easy to burn out an LED, and using a resistor is an excellent habit to get into. Many LED resistor calculators are available online if you aren't sure what value of resistor to use; I like the one at http://ledz.com/?p=zz.led.resistor.calculator.*

I2C Sensors

I2C, also referred to as *I-squared-C* or *Inter-Integrated Circuit*, has been called the serial protocol on steroids. It allows a large number of connected devices to communicate on a circuit, or *bus*, using only three wires: a data line, a clock line, and a ground line. One machine on the bus serves as the *master*, and the other devices are referred to as *slaves*. Each device is called a *node*, and each slave node has a 7-bit address, such as 0x77 or 0x43. When the master node needs to communicate with a particular slave node, it transmits a start bit, followed by the slave's address, on the data (SDA) line. The slave sees its address come across the data line and responds with an acknowledgment, while the other slaves go back to waiting to be called on. The master and slave then communicate with each other, using the clock line (SCL) to synchronize their communications, until all messages have been transmitted.

The Raspberry Pi has two GPIO pins, 3 and 5, that are preconfigured as the I2C protocol's SDA (data) and SCL (clock) pins. Any sensor that uses the I2C protocol can be connected to these pins to easily communicate with the Pi, which serves as the master node. If you end up using more than one I2C sensor on your rover, you may find it helpful to add another small breadboard to the rover, with the two power rails running down the side of the board being used for the data and clock lines rather than power.

To use the I2C protocol on the Pi, you first need to enable it by editing a few system files. Start with `sudo nano /etc/modules` and add the following lines to the end of the file:

```
i2c-bcm2708
i2c-dev
```

Next, install the I2C utilities with the following:

```
sudo apt-get install python-smbus
sudo apt-get install i2c-tools
```

Finally, you may need to edit your *blacklist* file, if you have one. (Remember, if it exists, you can find it in */etc/modprobe.d/raspi-blacklist.conf*.) If you have one, comment out the following two lines by adding a hashtag to the beginning of each line:

```
#blacklist spi-bcm2708
#blacklist i2c-bcm2708
```

Save the file, then reboot your Pi with the following, and you should be ready to use the I2C protocol with your sensors:

```
sudo shutdown -r now
```

To see if everything installed correctly, try running the `i2cdetect` utility:

```
sudo i2cdetect -y 1
```

It should bring up the screen in Figure 10-6.

Obviously, no devices are showing up in the illustration because we haven't plugged any in yet, but the tool is working. If by chance you have devices plugged in but they don't show up, or if the tool fails to start at all, instead try typing the following:

```
sudo i2cdetect -y 0
```

The `1` or `0` flag depends on the Pi revision you happen to have. If you have a Revision 1 board, you'll be using the `0` flag; Revision 2 owners will need to use the `1`.

```
pi@raspberrypi ~/MAKE $ sudo i2cdetect -y 1
     0  1  2  3  4  5  6  7  8  9  a  b  c  d  e  f
00:          -- -- -- -- -- -- -- -- -- -- -- -- --
10: -- -- -- -- -- -- -- -- -- -- -- -- -- -- -- --
20: -- -- -- -- -- -- -- -- -- -- -- -- -- -- -- --
30: -- -- -- -- -- -- -- -- -- -- -- -- -- -- -- --
40: -- -- -- -- -- -- -- -- -- -- -- -- -- -- -- --
50: -- -- -- -- -- -- -- -- -- -- -- -- -- -- -- --
60: -- -- -- -- -- -- -- -- -- -- -- -- -- -- -- --
70: -- -- -- -- -- -- -- --
pi@raspberrypi ~/MAKE $
```

Figure 10-6. *The i2cdetect utility*

To test the `i2cdetect` utility, connect an I2C sensor to your Pi, such as the digital compass —the HMC5883L. Connect the compass's Vcc and GND pins to the Pi's 2 and 6 pins, and then the SDA and SCL pins to pins 3 and 5, respectively. When you start the `i2cdetect` utility, you should now see the results in Figure 10-7, which shows the compass's (preconfigured) 12C address of `0x1e`.

```
pi@raspberrypi ~/MAKE $ sudo i2cdetect -y 1
     0  1  2  3  4  5  6  7  8  9  a  b  c  d  e  f
00:          -- -- -- -- -- -- -- -- -- -- -- -- --
10: -- -- -- -- -- -- -- -- -- -- -- -- -- -- 1e --
20: -- -- -- -- -- -- -- -- -- -- -- -- -- -- -- --
30: -- -- -- -- -- -- -- -- -- -- -- -- -- -- -- --
40: -- -- -- -- -- -- -- -- -- -- -- -- -- -- -- --
50: -- -- -- -- -- -- -- -- -- -- -- -- -- -- -- --
60: -- -- -- -- -- -- -- -- -- -- -- -- -- -- -- --
70: -- -- -- -- -- -- -- --
pi@raspberrypi ~/MAKE $
```

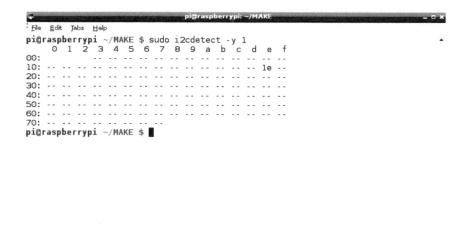

Figure 10-7. *i2cdetect showing connected HMC5883L*

HMC5883L Compass

You have it connected now, so let's configure the compass. The following script sets up the I2C bus, and after reading the values from the compass, performs a little bit of math wizardry to translate its readings into a format that you and I are used to:

```
import smbus
import math

bus = smbus.SMBus(0)
address = 0x1e

def read_byte(adr):
   return bus.read_byte_data(address, adr)

def read_word(adr):
   high = bus.read_byte_data(address, adr)
   low = bus.read_byte_data(address, adr+1)
   val = (high << 8) + low
   return val

def read_word_2c(adr):
   val = read_word(adr)
   if val >= 0x8000:
       return -((65535 - val) + 1)
   else:
       return val

def write_byte(adr, value):
   bus.write_byte_data(address, adr, value)

write_byte (0, 0b01110000)
write_byte (1, 0b00100000)
write_byte (2, 0b00000000)

scale = 0.92
x_offset = -39
y_offset = -100

x_out = (read_word_2c(3) - x_offset) * scale
y_out = (read_word_2c(7) - y_offset) * scale

bearing = math.atan2(y_out, x_out)
if bearing < 0:
   bearing += 2 * math.pi
print "Bearing: ", math.degrees(bearing)
```

Here, after importing the necessary libraries (*smbus* and *math*), we define functions (`read_byte()`, `read_word()`, `read_word2c()`, and `write_byte()`) to read from and write values (either single bytes or 8-bit values) to the sensor's I2C address. The three `write_byte()` lines write the values 112, 32, and 0 to the sensor to configure it for reading. Those values are normally listed in a sensor's datasheet.

The script then reads the x-axis and y-axis values from the sensor and uses the *math* library's `atan2()` (inverse tangent) function to find the sensor's bearing. The `x_offset` and `y_offset` values are subject to change and are dependent on your current location on the Earth's surface; the easiest way to determine what they should be is to run the script with a working compass nearby to compare values. When you run the script, remember that the side of the chip with the soldered headers is the direction in which the board is "pointed." Compare the readings and tweak the values of the `x_offset` and `y_offset` values until the readings from the two compasses match. Now you can determine which direction your rover is headed. You shouldn't experience any interference from your Pi or from the motors on your magnetic sensor; the fields generated by those devices are too weak to make a difference in the sensor's readings.

As always, save this script in your rover's folder for addition to your main program.

BMP180P Barometer

The BMP180P barometer/pressure chip is another sensor that runs on the I2C protocol. Again, connect the SDA and SCL pins to either the Pi's 3 and 5 pins, or the rails on the breadboard if you've gone that route, and the GND pin to the Pi's 6 pin. This time, however, connect the Vcc pin to the Pi's 1 pin, *not* the 2 pin. This sensor needs only 3.3V, and powering it with 3.3V instead of 5V ensures that it will output only 3.3V and not damage the Pi's delicate GPIO pins. After you've connected everything, run the `i2cdetect` utility to make sure that you see the sensor's address, which should be `0x77`.

Like a few of the other sensors, this one needs an external library in order to work, and that library is available from Adafruit. In your terminal, make sure you're in the main rover folder and type the following:

```
wget http://bit.ly/NJZOTr
```

Rename the downloaded file with this command and the library is ready to use, as long as it's in the same folder as your script:

```
mv NJZOTr Adafruit_BMP085.py
```

You'll also need another script from Adafruit in the same directory, the `Adafruit_I2C` library. To get it, in a terminal enter the following:

```
wget http://bit.ly/1pHgMxF
```

and then rename it with the following:

```
mv 1pHgMxF Adafruit_I2C.py
```

Now you have both necessary libraries. To read from the sensor, create the following script in your rover folder to convert to Fahrenheit:

```
from Adafruit_BMP085 import BMP085
bmp = BMP085(0x77)
temp = bmp.readTemperature()
pressure = bmp.readPressure()
altitude = bmp.readAltitude()
```

```
    print "Temperature:  %.2f C" %temp
    print "Pressure:     %.2f hPa" %(pressure/100.0)
    print "Altitude:     %.2f" %altitude
```

The Adafruit library is nice because it handles all the intricacies of communicating over the I2C bus for us; all we have to do is call the functions `readTemperature()`, `readPressure()`, and `readAltitude()`. If you're not in one of the 99% of countries using Celsius for temperature, just add the following line:

```
temp = temp*9/5 + 32
```

Nintendo Wii Devices

You can also use the I2C library to communicate with other devices, of course; it's not unheard of to connect a Nintendo Wii nunchuk to the Pi with a special adapter, called a Wiichuck adapter (Figure 10-8).

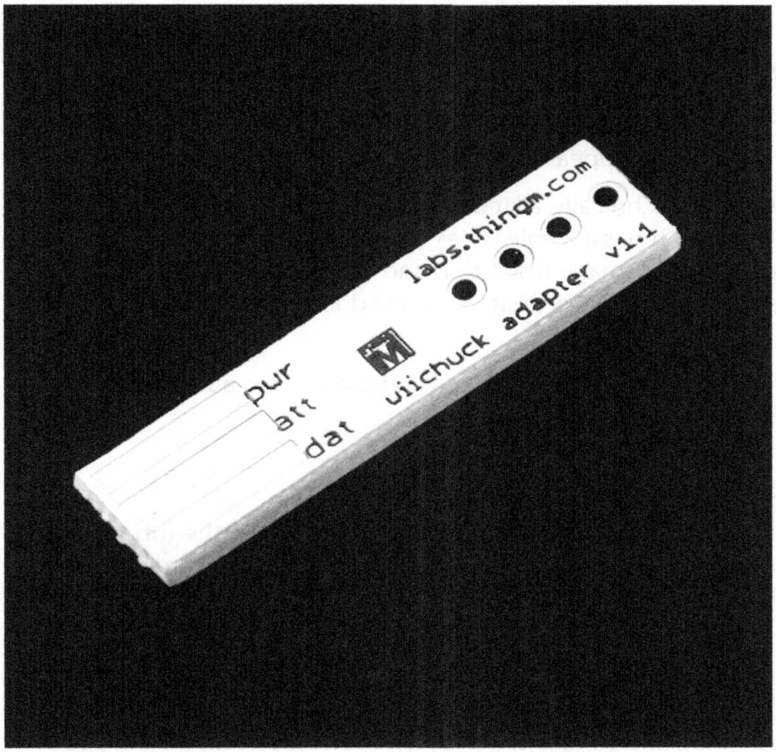

Figure 10-8. *Wiichuck adapter*

You can then read the values from the nunchuk's joystick, buttons, and onboard accelerometer to control things like motors, cameras, and other parts of the robot.

Camera

The last thing we need to go over is the Pi's camera; it *is* technically a sensor, and you can use it to take pictures of your rover's surroundings and even stream a live feed over the local network and navigate that way.

Hooking up the camera is fairly straightforward. If you're sticking with the flex cable that came with the camera, you're almost finished already. Insert the flex cable into the small connector between the Ethernet port and the HDMI connector. To insert it, you may have to pull up slightly on the tabs on both sides of the connector. Insert the cable with the silver connections facing the HDMI port, as far as it will go, and then press down on the edges of the connector to lock it into place.

If you're using an extension cable such as the one from BitWizard (*http://bit.ly/1xqiybL*), follow their instructions as to hooking up the flex and the ribbon cables. When you're finished connecting the camera, enable it in the *raspi-config* file if you haven't done so already by typing the following and enabling it there (option 5):

```
sudo raspi-config
```

Once enabled, to test the camera, open a terminal window and type:

```
raspistill -o image.jpg
```

After a short pause, *image.jpg* should appear in the current directory.

Raspistill is an amazing program. Technically, all it does is take still pictures with the Raspberry Pi camera module. In reality, it has a whole series of options, including the ability to take time-lapse sequences, to adjust image resolution and image size, and so forth. Play around with the flags listed on the Raspberry Pi site's camera documentation page (*http://bit.ly/1CLWpWL*).

To use the Python library now available for the camera (Python 2.7 and above), enter the following in your terminal to install it:

```
sudo apt-get update
sudo apt-get install python-picamera
```

You're now ready to use the camera. If you plan to place the camera in the robotic arm attachment, refer back to Chapter 7 as to how to mount it there. Then you can use it with the Python library with a script such as this:

```
import picamera
camera = picamera.PiCamera()
camera.capture('image.jpg')
```

This will simply capture *image.jpg* and store it in the local directory. One nice thing about the Python library as opposed to the command-line interface is that the default image size for the Python module is much smaller than the command-line default.

If you would like to record video with the camera, it's as simple as this:

```
import picamera
import time
camera = picamera.PiCamera()
camera.start_recording('video.h264')
time.sleep(5)
camera.stop_recording()
```

This will record for 5 seconds and then stop.

Live camera feed

All of these are nice if you simply want to travel to a point and then take pictures or video after you arrive. But what if you would like to navigate using the feed from the camera? This, too, is possible, by streaming the video feed from the camera over the local ad hoc network you've set up and playing the stream on the computer you're using to remotely control the Pi. To do this, you'll need the VLC media player installed on both the Pi and your control computer. On the Pi, it's a simple:

```
sudo apt-get install vlc
```

to install it; on your controlling computer, VLC is available for Linux, Windows, and OS X.

The stream will be broadcast using Real Time Streaming Protocol (RTSP). This protocol is a common network video-streaming interface, and VLC is easy to set up to both transmit and receive and decode it. Once VLC is installed on the Pi, start the stream with the following:

```
raspivid -o - -t 0 -n -w 600 -h 400 -fps 12 |
cvlc -vvv stream:///dev/stdin --sout
'#rtp{sdp=rtsp://:8554/}' :demux=h264
```

Then move to your control computer, open VLC, and open a network stream from rtsp://<Your Pi IP>:8554. It's a small, 600 x 400 window, so not too much bandwidth should be needed. There's also likely to be a delay of several seconds, so this may not be an optimal way of controlling your rover in a situation where fast response times are important.

You may run into the problem of your Pi shutting down as soon as you issue the streaming command shown here. It seems that the command draws a lot of power—sometimes, enough to shut everything off. If that does happen, try a different power supply (if you're powering the Pi from a USB wall charger) or a different battery pack (if you're using batteries such as the Li-Poly battery pack). Experimentation is always helpful; I had success simply by using a shorter USB power cable at one point.

If you can't *get it to work, your particular Pi/power/VLC combination may just be too ill-suited for live streaming video. In that case, you'll just have to remain in view of your rover to control it—which is not the end of the world.*

That by no means covers all of the sensors that are available for your rover, but it should give you a pretty good start. Many sensors are just switches at heart, and if not, there may be a library available to read from them. Or they may follow the I2C protocol, making them easy to add to your rover's sensor network. In the next chapter, we'll cover putting all of these snippets of code together and controlling (and reading from) the rover by using one program.

Final Code and Conclusion

11

At this point in the build, you should have at least two things: a working rover, and a directory on the Pi full of small Python scripts that do all kinds of neat things separately, but don't work together very well. As the last part of this build, we'll have to combine all of the scripts into one large working program that does everything we expect it to.

When setting up this program, we need to ask ourselves a few questions. First, how many sensors are we going to be using? There are a lot of different ones, and you have only a limited number of GPIO pins available (unless they're all on the I2C bus). Also, some of the sensors don't play well together in the same program. For instance, the SHT15 temperature sensor works great when you test it by itself. But when you try to use another GPIO pin as an output (such as for a motor), the `sht1x` library re-declares the GPIO pin setup, which negates all the setup you do in the main script. Not to worry, of course—we can use the `getTemperature()` function from the BMP180 sensor to get the ambient temperature.

So once you've figured out what sensors you're going to use, and where to place them, and how to wire them, all that remains is to write the final program. Simple, right? Actually, as I said at the beginning of the book, if you've been following along, most of this work is already done —all you're doing is putting the pieces together.

I designed the program to be interactive. There's no robotic autonomy in this program, as such a program would probably end up being thousands of lines of code. Rather, the program displays all pertinent sensor data (temperature, pressure, bearing, location, etc.) and then asks for input from the user as to what to do next. If you tell the rover to move forward, the program calls the `moveForward()` function. Then the sensors are polled, the sensor data is displayed, and the user is prompted again. The `moveForward()` function continues to execute until the user stops it. This allows the rover to continue to move until you tell it otherwise, rather than moving forward a few feet and then stopping and waiting for further instructions.

If you choose to build on this script and pursue autonomous behavior, your best bet is probably to go through a continuous loop, polling the sensors for data one by one. You can then put interrupts related to each sensor that tell the rover what to do if any of the sensors read a particular value.

Perhaps the most obvious use of this algorithm is to make the rover follow a preprogrammed route using GPS waypoints. By calling `gpsd.fix.latitude` and `gpsd.fix.longitude`, you can determine whether the rover should move forward, turn left, or turn right. When those values match your first set of coordinates, you can execute a preplanned action and then continue. To give you an example, using the GPS sensor *only*:

```
# Destination: 36.21 degrees N, 116.53 degrees W.
# When we reach that point, turn due south and
# continue driving
# Assume that we're approaching from the West


while True:
    curLat = gpsd.fix.latitude
    curLon = gpsd.fix.longitude
    # Longitude degrees W are delineated with a
    # negative sign in NMEA strings
    if (curLat == 36.21) and (curLon == -116.53):
        allStop()
        spinLeft()
        time.sleep(1)
        allStop()
        takePicture()
    elif curLat < 36.21:
        # You've gone too far,
        # back up
        moveBackward()
        continue
    elif curLat > 36.21:
        # You're not there yet
        # Keep going
        moveForward()
        continue
```

This is by no means a complete section of code, has *not* been tested, and barely covers any of the possible actions based on the rover's location, but it should give you some ideas, both of what is possible with a rover and what is required to achieve it. Programming an "intelligent" rover or robot is no small task, as you need to try to plan for all possible situations and then program responses to those situations—at least until artificial intelligence makes some serious strides forward. That, in fact, is part of the fun of robotic programming: not only trying to anticipate all possible situations, but also trying to program behaviors and algorithms so that the rover can react to those situations, as well as those that you (inevitably) didn't think of or plan for. This makes what I call a *robust* program; should something unexpected arise, the rover has a default behavior that it can fall back on that will always work.

In the meantime, you'll be driving your rover with your laptop, calling functions by pressing keys. I've tried to keep to the traditional gaming keymap: `W` to move forward, `Z` to move backward, A and D to move left and right, respectively, and S to stop. In addition, there are keys to raise the arm, lower the arm, and take a picture. In each case, the program asks for user input, calls the required function, waits a second, polls the sensors, clears the screen, and displays the sensor readings and the input prompt again. Before everything else starts, the program asks the user if a GPS is connected; if you don't have one connected, the program skips the location query. This is done because if the program tries to query a nonexistent GPS module, it will break.

```
# This script runs the rover,
# displaying readings from the sensors
# every time it gets input
# from the user

import time
import os
import RPi.GPIO as GPIO
import subprocess
from sht1x.Sht1x import Sht1x as SHT1x
import smbus
import math
from gps import *
import threading
from Adafruit_BMP085 import BMP085

GPIO.setwarnings(False)
GPIO.setmode(GPIO.BOARD)
GPIO.setwarnings (False)

# Motor setup
# 19 = IN1
# 21 = ENA
# 23 = IN2
GPIO.setup(19, GPIO.OUT)
GPIO.setup(21, GPIO.OUT)
GPIO.setup(23, GPIO.OUT)
# 22 = IN3
# 24 = ENB
# 26 = IN4
GPIO.setup(22, GPIO.OUT)
GPIO.setup(24, GPIO.OUT)
GPIO.setup(26, GPIO.OUT)

def moveForward():
    # r forward
    GPIO.output(21, 1)
    GPIO.output(19, 0)
    GPIO.output(23, 1)
    # l forward
    GPIO.output(24, 1)
    GPIO.output(22, 0)
    GPIO.output(26, 1)

def moveBackward():
    # r backward
```

```
        GPIO.output(21, 1)
        GPIO.output(19, 1)
        GPIO.output(23, 0)
        # l backward
        GPIO.output(24, 1)
        GPIO.output(22, 1)
        GPIO.output(26, 0)

    def allStop():
        GPIO.output(21, 0)
        GPIO.output(24, 0)

    def spinRight():
        # leftforward, rightbackward
        GPIO.output(24, 1)
        GPIO.output(22, 0)
        GPIO.output(26, 1)
        GPIO.output(21, 1)
        GPIO.output(19, 1)
        GPIO.output(23, 0)

    def spinLeft():
        # rightforward, leftbackward
        GPIO.output(21, 1)
        GPIO.output(19, 0)
        GPIO.output(23, 1)
        GPIO.output(24, 1)
        GPIO.output(22, 1)
        GPIO.output(26, 0)

    # GPS setup
    gpsd = None
    class GpsPoller(threading.Thread):
        def __init__(self):
            threading.Thread.__init__(self)
            global gpsd
            gpsd = gps(mode=WATCH_ENABLE)
            self.current_value=None
            self.running = True
        def run(self):
            global gpsd
            while gpsp.running:
                gpsd.next()

    # Compass setup
    bus = smbus.SMBus(0)
    compAddress = 0x1e
    def read_byte(adr):
        return bus.read_byte_data(compAddress, adr)

    def read_word(adr):
        high = bus.read_byte_data(compAddress, adr)
        low = bus.read_byte_data(compAddress, adr+1)
        val = (high << 8) + low
```

```
    return val

def read_word_2c(adr):
    val = read_word(adr)
    if val >= 0x8000:
        return -((65535 - val) + 1)
    else:
        return val

def write_byte(adr, value):
    bus.write_byte_data(compAddress, adr, value)

def getBearing():
    write_byte(0, 0b01110000)
    write_byte(1, 0b00100000)
    write_byte(2, 0b00000000)
    scale = 0.92
    x_offset = -39
    y_offset = -100
    x_out = (read_word_2c(3) - x_offset) * scale
    y_out = (read_word_2c(7) - y_offset) * scale
    bearing = math.atan2(y_out, x_out)
    if bearing < 0:
        bearing + 2 * math.pi
    return str(math.degrees(bearing))


# Robotic arm servo setup

def liftArm():
    for i in range(50, 90):
        subprocess.call("echo 2=" + str(i) + "/dev/servoblaster", shell=True)
        time.sleep(0.5)

def lowerArm():
    for i in reversed(range(50, 90)):
        subprocess.call("echo 2=" + str(i) + "/dev/servoblaster", shell=True)
        time.sleep(0.5)

# Rangefinder setup
GPIO.setup(15, GPIO.OUT)
GPIO.setup(13, GPIO.IN)

def getRange():
    time.sleep(0.3)
    GPIO.output(15, 1)
    time.sleep(0.00001)
    GPIO.output(15, 0)
    while GPIO.input(13) == 0:
        signaloff = time.time()
    while GPIO.input(13) == 1:
        signalon = time.time()
    timepassed = signalon - signaloff
    distance = timepassed * 17000
```

```
        return str(distance)

    # Pressure and temperature
    bmp = BMP085(0x77)
    def getTemperature():
        return str(bmp.readTemperature())

    def getPressure():
        return str(bmp.readPressure()/1000)

    if __name__ == '__main__':
        gpsQuery = raw_input("Do you have a GPS connected? (y/n) ")
        if gpsQuery == 'y':
            gpsp = GpsPoller()
            try:
                gpsp.start()
                while True:

                    # Get command from user
                    os.system("clear")
                    print "Range to target: " + getRange()
                    print "Temp: " + getTemperature() + "C"
                    print "Pressure: " + getPressure() + "kPa"
                    print "Location: " + str(gpsd.fix.longitude)
                    + ", " + str(gpsd.fix.latitude)
                    print "Bearing: " + getBearing() + " degrees"
                    print "W = forward"
                    print "Z = backward"
                    print "A = left"
                    print "D = right"
                    print "S = stop"
                    print "O = raise arm"
                    print "P = lower arm"
                    print "I = take picture"

                    command = raw_input("Enter command(Q to quit): ")
                    if command == "w":
                        moveForward()
                        time.sleep(0.5)
                        continue
                    elif command == "z":
                        moveBackward()
                        time.sleep(0.5)
                        continue
                    elif command == "a":
                        spinLeft()
                        time.sleep(0.5)
                        continue
                    elif command == "d":
                        spinRight()
                        time.sleep(0.5)
                        continue
                    elif command == "s":
                        allStop()
```

```
                time.sleep(0.5)
                continue
            elif command == "o":
                liftArm()
                time.sleep(0.5)
                continue
            elif command == "p":
                lowerArm()
                time.sleep(0.5)
                continue
            elif command == "i":
                subprocess.call("raspistill -o image.jpg",
                shell=True)
                time.sleep(0.5)
                continue
            elif command == "q":
                gpsp.running=False
                gpsp.join()
                GPIO.cleanup()
                break
            else:
                print "Command not recognized. Try again."
                time.sleep(1)
                continue
    except (KeyboardInterrupt, SystemExit):
        gpsp.running = False
        gpsp.join()
        GPIO.cleanup()
else:
    try:
        while True:
            # Get command from user
            os.system("clear")
            print "Range to target: " + getRange()
            print "Temp: " + getTemperature() + "C"
            print "Pressure: " + getPressure() + "kPa"
            print "Bearing: " + getBearing() + " degrees"
            print "W = forward"
            print "Z = backward"
            print "A = left"
            print "D = right"
            print "S = stop"
            print "O = raise arm"
            print "P = lower arm"
            print "I = take picture"

            command = raw_input("Enter command
            (Q to quit): ")
            if command == "w":
                moveForward()
                time.sleep(0.5)
                continue
            elif command == "z":
                moveBackward()
```

```
                time.sleep(0.5)
                continue
            elif command == "a":
                spinLeft()
                time.sleep(0.5)
                continue
            elif command == "d":
                spinRight()
                time.sleep(0.5)
                continue
            elif command == "s":
                allStop()
                time.sleep(0.5)
                continue
            elif command == "o":
                liftArm()
                time.sleep(0.5)
                continue
            elif command == "p":
                lowerArm()
                time.sleep(0.5)
                continue
            elif command == "i":
                subprocess.call("raspistill -o
                image.jpg",
                shell=True)
                time.sleep(0.5)
                continue
            elif command == "q":
                gpsp.running=False
                gpsp.join()
                GPIO.cleanup()
                break
            else:
                print "Command not recognized.
                Try again."
                time.sleep(1)
                continue
    except (KeyboardInterrupt, SystemExit):
        GPIO.cleanup()
```

Now, keep in mind that this is only a starter script, and it doesn't use all of the sensors we went over in Chapter 10. Because your Pi has a limited number of GPIO pins, you may have to play around with power rails on your breadboard, adding and subtracting I2C devices, and other ways of managing the sensors on your rover. When it's running, you'll see a command window like that in Figure 11-1.

```
Temp: 25.1C
Pressure: 100kPa
Location: -149.727843333, 61.198966667
Bearing: 42.0
W = forward
Z = backward
A = left
D = right
S = stop
O = raise arm
P = lower arm
I = take picture
Enter command (Q to quit): █
```

Figure 11-1. *Rover command program window*

You may even decide to design a GUI for your rover interface. If you decide to do that, I suggest researching Python's *Tkinter* library (*https://wiki.python.org/moin/TkInter*). It's functional more than fashionable, but its small learning curve makes it possible to design fully working user interfaces for your Python scripts.

Whatever you decide to do with your rover, above all have fun with it. Once you've solved the problems inherent in designing a rover from scratch, you can modify and tweak it to your heart's content.

I look forward to seeing what you come up with!

A Setting Up the Pi

You've probably noticed that after you get the Pi up and running, working with it is pretty straightforward. You have root access (using `sudo`) to any files you need to change, such as */etc/network/interfaces*, or the */etc/rc.local* file. You can plug in your keyboard, mouse, and monitor, and work with it as a standard desktop machine, or (as I prefer) you can just hook it to your network and remotely log into it via SSH (or VNC, if you need a graphic desktop environment). With the wireless working, you can put the Pi in your rover and do all of your programming work without taking it out of the robot.

But what about setting it up in the first place? Even if you bought your Pi with the now-available NOOBS-preloaded SD card, you still need to install Raspbian and set it up to be easily accessed remotely. If you just bought yourself a bare-bones setup and need to get NOOBS, you may be a little confused. Sure, there are instructions on the raspberrypi.org website, but who reads instructions? And if you're new to this whole Raspberry Pi thing, you may need some help.

That is what this appendix is for. (The word *appendix* comes from the Latin word *appendere*, meaning "to hang upon," or "to explain OS installation.") Let's quickly go through the process of downloading NOOBS, formatting your SD card, installing Raspbian, and working through the `raspi-config` tool. If you bought a preformatted card (always a good idea, as it's only about $7, and all proceeds go to the nonprofit Raspberry Pi Foundation), you can skip ahead to the installation section.

And why download NOOBS, you ask? NOOBS is handy because it contains all the files you need in one easy download. In fact, it contains the installation files for several operating systems, including Raspbian and Kodi Entertainment Center, so should you want to experiment with other OSs on your Pi, by downloading NOOBS you have access to a virtual cornucopia of operating system goodness.

Download NOOBS

NOOBS, which stands for *New Out Of Box Software*, is available for free from *http://www.raspberrypi.org/downloads*. The version as of this writing is 1.3.5. On that page, click the ZIP file to begin the download and get a cup of coffee (or go to bed—it's a hefty 1.3GB download).

After the ZIP file is downloaded, extract the files. You should end up with a folder containing files similar to what you see in Figure A-1.

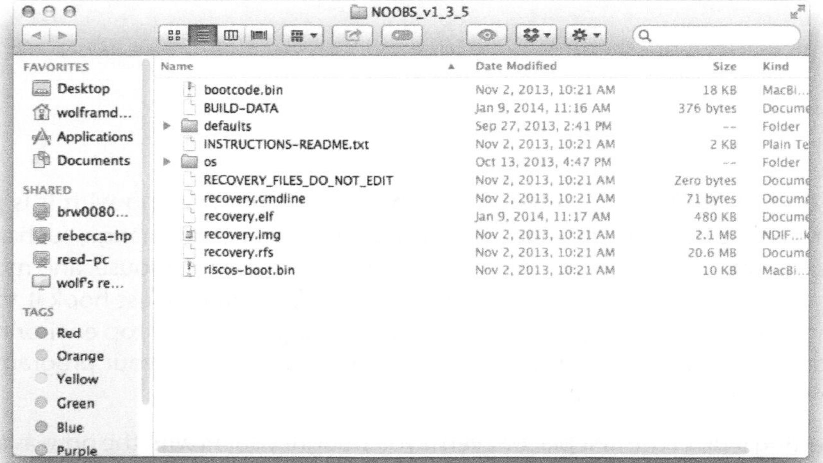

Figure A-1. *Contents of the NOOBS folder*

Set that folder aside for a moment. Before we do anything with it, we need to get the SD card ready.

Download the SD Card Formatting Tool

Although you can *probably* just drag and drop the NOOBS files onto a blank SD card and expect it to work, you may not want to take that chance. Depending on your computer's operating system and the format currently on the card (which often depends on the card's manufacturer), the dragged-and-dropped files may or may not install correctly onto the card. I have had luck with a Mac and a PNY card, but the same card did not accept the files when I attempted to drag and drop them from a Windows installation. To avoid any guesswork, I highly recommend getting the card-formatting tool from the SD Association's website.

Point your browser to *http://www.sdcard.org/downloads/formatter_4/*. On the left side of the page, you'll see your download choices (Figure A-2). Select your operating system, agree to the terms, and download the tool. Unlike the NOOBS file, it's quite small.

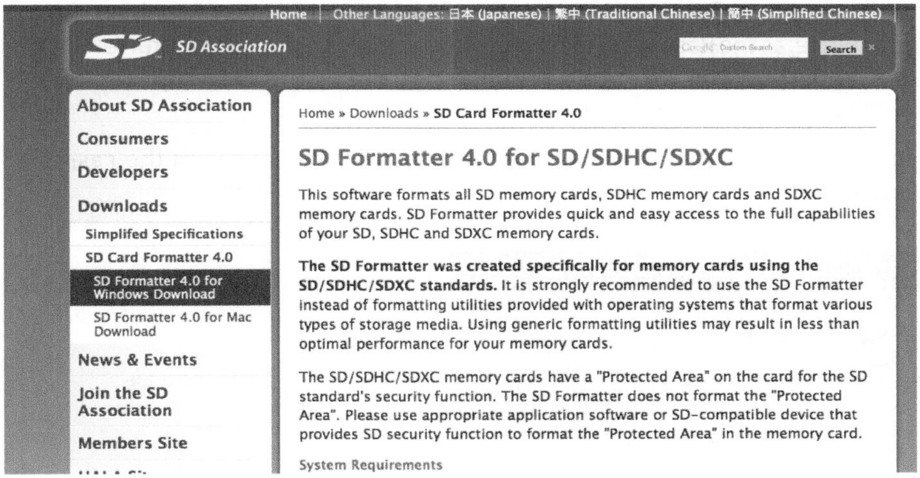

Figure A-2. *The NOOBS download*

When it is downloaded, open up the file and follow the instructions to install the SD card formatting tool on your machine.

Format and Fill Your Card

While we're on the subject of cards, which one are you using? You'll need at least 8GB for your install. (The raspberrypi.org site says 4GB, but my experience is that 4GB isn't big enough.) Go larger if you like; there's no upper limit, as far as I know. I tend to go with 16GB.

Put the card in your computer's card slot. Open the formatting tool you installed earlier and follow the instructions to format the card.

> *Make sure you choose the correct drive to format. The tool will erase whatever disk you point it at! That includes your hard drive. Be careful!*

When the tool is done and the card is ready, just copy the files from the *NOOBS_v1_3_5* folder onto the card. That's it!

The raspi-config Tool

When the Pi starts up for the first time, you'll see a splash screen, giving you the option of which OS to install. Choose Raspbian (the first option) and click Install. The NOOBS tool will expand the file system to fill your SD card and then do a clean install of Raspbian. Feel free to watch for a while, as the Raspberry Pi Foundation has included some helpful reading material to watch while you wait. And you *will* wait: especially if you have a high-capacity SD card, the installation may take up to an hour.

When the install is done, you'll reboot the Pi, and then will be greeted with the `raspi-config` tool (Figure A-3).

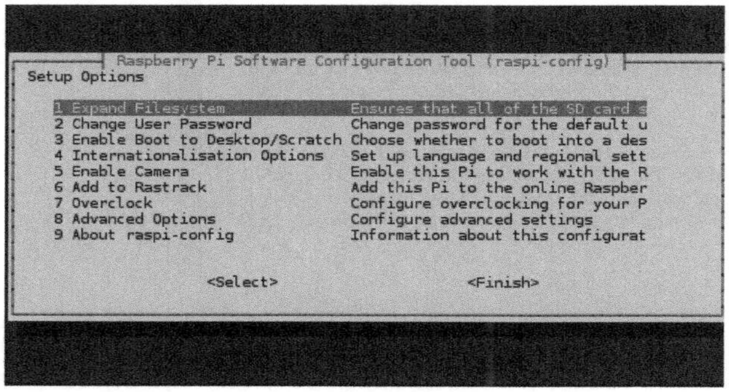

Figure A-3. *The raspi-config tool*

Getting around the tool is easy: use the up and down arrows to choose your line item, and then the right arrow and Enter key to select.

You used the NOOBS tool, so you can disregard the first item, Expand Filesystem, because NOOBS automatically expands the installation to fill the SD card. The second menu item lets you change the default username and password from `pi` and `raspberry`, respectively, to something more appropriate, should you so desire. It's probably unnecessary, unless you plan on exposing your Pi to the outside world via an unprotected network.

The third option allows you to choose whether you want to boot to a desktop environment, a command line, or the Scratch programming language IDE. If you choose to boot to a command line, you can always start a desktop by typing **`startx`** at the prompt.

The fourth option, Internationalisation Options, is important if you're *not* using the Pi in the United Kingdom. Work your way through the menus, choosing your locale, time zone, and the sort of keyboard you have. The locale menu is a little different to get around in, and can be a bit confusing. First of all, it's comprehensive, meaning that a *lot* of regions are listed, from Antigua to Zimbabwe. To choose your locale, move up and down with the arrow keys. The locale(s) currently selected will have an asterisk in the brackets (Figure A-4). To clear or add an asterisk, press the space bar while you have that line selected. According to the configuration instructions on the screen, when faced with a choice, choose the UTF-8 locale for your country.

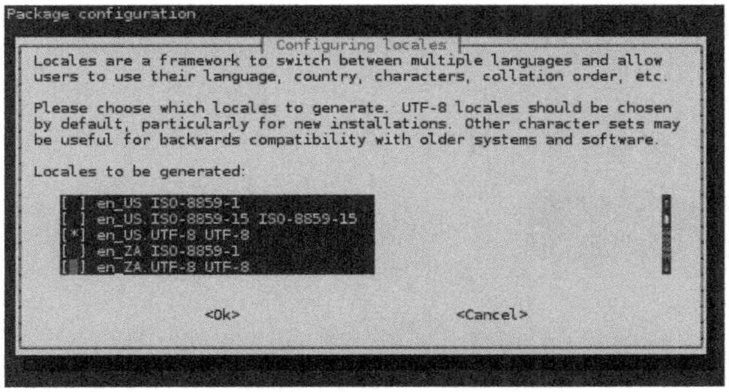

Figure A-4. *en_US locale selected in raspi-config*

When you're finished, press the Tab key to get to the <Ok> option, and press Enter. Follow that up by choosing your time zone (your system clock will be updated via a network time server if you have your Pi connected to the Internet while you run the tool) and your keyboard layout. In general, unless you have a super-duper high-tech keyboard, you can probably just agree with the defaults presented to you in the keyboard selection screens and press <Ok>. You'll know you got it right if, after you're finished, pressing Shift+2 gives you the result you expect—either an *at* sign (@) or a double quote (").

Enable Camera, the fifth option, does just what it says—configuring the Pi to work with the camera board. If you don't enable it, your Pi won't be able to work with the camera, and if you're planning on putting the IR camera in the robotic arm as per the rover design, you'll need camera support. Even if you're not going to use the Pi camera, I recommend enabling it. It doesn't cost you anything.

Option six, Add to Rastrack, simply adds your Pi to the global database/map of Raspberry Pis. Feel free to enable it if you're not feeling particularly paranoid today.

The seventh option, Overclock, gives you the ability to upgrade your 700MHz chip all the way to a screaming-fast 1GHz. This is totally up to you; I find it's not really necessary unless you're planning on doing a lot of intensive computing or working with video. It can make your Pi run a little hot, and *can* cause system instability. Experiment if you feel you must.

Option eight, Advanced Options, is important if only because of its fourth submenu item, SSH (Figure A-5). It's important that you enable the SSH server on your Pi, so you can log in remotely and work on it while it's installed in your rover or other project. Experiment with the others as you like.

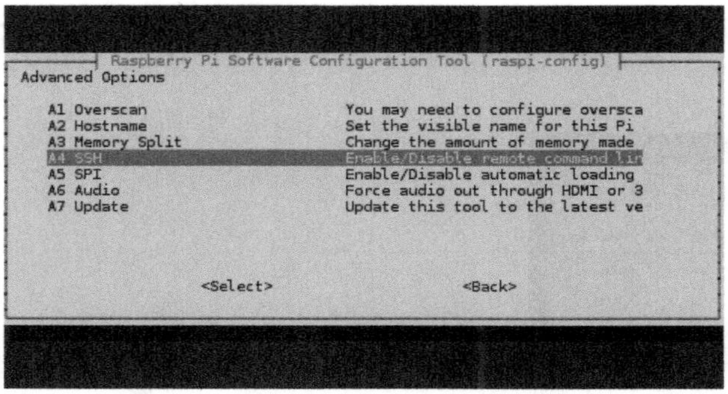

Figure A-5. *Enabling the SSH tool*

When you're finished playing with the options in the tool, select Finish and reboot your Pi if necessary.

The final part of the setup process is to update your Pi. Updates are continually being released, and it's quite likely that one or more of the packages on your Pi have been updated since the NOOBS tool was released. Open a terminal, and at the prompt type **`sudo apt-get update`** and **`sudo apt-get upgrade`**. That will make sure all of your installed software is cutting-edge. Depending on the number of updates to the software you have installed, this update and upgrade process can take quite a while—not quite as long as the initial installation, but still long enough to go enjoy a nice cup of tea.

That is a short-and-sweet guide to getting your Pi ready to program and build. As you progress through the build, you may find yourself downloading other software packages as well; it's not uncommon to end up with an SD card that's completely tailored to one project in terms of installed software and written programs and scripts. This is another advantage to the Pi's hard drive system: you can set that SD card aside, buy another one for under $30, and start fresh on the next project. It's not like buying another 1TB hard drive for your laptop every time you start a new build.

Now that you're up and running, you may want to peek at Appendix B if you need a refresher on (or an introduction to) the Python scripting language.

B Intro to Python

As you've probably noticed from flipping through the book, all of my scripts for the robot are written in Python. There are some good reasons for this. If you are already a skilled Python aficionado, you can probably skip this portion of the book. On the other hand, if you need a refresher, or if Python is completely new to you, read on for a quick-and-dirty introduction to this powerful language.

Python on the Pi: A History in Four Paragraphs

Python was born in 1989 to Guido van Rossum, who had a crazy idea that programming should be accessible to everybody, not just geeks with broken glasses and pocket protectors. He was working with a language called ABC, and wanted a language that would fix some of its problems and add new features. The result, created over a Christmas holiday, was Python. He continues to have a hand in the development of the language, and has been gifted by the Python community with the title Benevolent Dictator for Life, or BDFL.

Contrary to what some believe, Python is *not* named after the snake. Rather, it is named for the British comedy troupe Monty Python, of whom van Rossum is a huge fan. (If you have never seen a Monty Python sketch, you should go to YouTube *right now* and watch a few. I recommend the Dead Parrot sketch, the Ministry of Silly Walks, and the Argument Clinic.) No, you don't need to be a fan to use the language, but it helps you get some of the in-jokes.

Because of its Monty Python origins, you'll find references to the comedy scattered throughout Python. Instead of the common `foo` and `bar` example functions and variable names, you'll find `spam` and `eggs`. "Knights of Ni" references abound in tutorials and books. Even the integrated development environment (IDE) called IDLE is named after one of the members (see if you can guess which one).

As to Python on the Pi, the Pi's creators (Eben Upton, Rob Mullins, Jack Lang, and Alan Mycroft) wanted a small, cheap computer that anybody could learn to program on. Knowing how simple and how powerful Python was, they included it as the default language on the Pi. Yes, you can program its ARM processor in C or even (if you're particularly masochistic) assembly language, but after learning about Python, why would you want to?

Using IDLE

Perhaps the best way to get an introduction to the language is by using its real-time development environment, IDLE. On your Raspberry Pi desktop, doubleclick the IDLE icon (Figure B-1).

Figure B-1. *The IDLE icon*

This opens an interactive screen, as you see in Figure B-2.

Let's start with the first program any programmer ever learns. In your IDLE prompt, type the following:

```
print "Hello, world!"
```

You should be rewarded with:

```
'Hello, world!'
```

If you've programmed in different languages, you should immediately notice a difference. To print "Hello, world!" in C++, you'd need to type the following:

```
# include <iostream>
using namespace std;
int main()
{
    cout << "Hello, world!" << endl;
    return 0;
}
```

Figure B-2. *The IDLE environment*

Python takes only one line to do the same thing. There's also a noticeable lack of semicolons, and opening and closing braces. Python uses indentations and blank space to delineate blocks of code. If you need to block out an `if` statement, for example, you end the statement with a colon (`:`), and then the conditional statements are all indented. When the conditional statements end, the indentation ends.

Lines don't end with semicolons; rather, when a line is over, it's just over. So to illustrate an `if` block, for example:

```
if x < y:
  print "x is less than y"
  print "This block of code is now over"

print "This is a new code section."
```

This has the effect of making Python code much easier to read, and much easier to debug.

As you progress in your programming skills, get in the excellent habit of commenting your code; it makes it easier not only for others to understand it, but also for you to understand your own code when you go back to it after several months of doing something completely different.

Now, back in IDLE, type **`x = 4`**.

Then type **`x`** and press Enter. You'll be rewarded with the following:

```
4
```

You've just defined the variable x as an integer: a variable that can hold a small number (well, a number between 0 and 65533).

Type **`y = "This is a string"`**. Then type **`y`** and press Enter. You'll be rewarded with the following:

```
'This is a string'
```

You've just defined the variable y as a string, a collection of characters.

Now, type **`x + 4`** and press Enter. You'll see:

```
8
```

Type **`y + " and is long"`** and press Enter, and you'll see this:

```
'This is a string and is long'
```

To finish off this little lesson, type **`x +" is not a string"`** and press Enter. You'll see your first error:

```
TypeError: unsupported operand type(s) for +:'int' and 'str'
```

This illustrates that Python is a *dynamically typed* language. You don't need to tell it that x is an integer; it knows from your previous use of x. It also knows that y is a string, and that you can't add a string and an integer to each other. However, if you were to convert the integer to a string, with `str(x) + " is not a string"`, the result of that command would be:

```
'4 is not a string'
```

I've mentioned integers and strings. There are also `long` variables, ones that can hold large numbers (519234L, for example); `floats`, which can hold what we think of as fractional, decimal, or *real* numbers (1.2345); and `complex`, which can hold what mathematicians also refer to as *imaginary* numbers, that contain the square root of –1. This is often written as `i` (or in engineering as *j*). So, for example, 3.14 times the square root of –1 will be displayed as `3.14j` in Python. Python's standard library allows you to perform all of the standard operations with those numbers: exponentiation, multiplication, division, addition, and subtraction. If you wish to extend your capabilities, you can import the `math` module for additional functions like `floor`, `ceil`, and others.

Python's other main data types are lists, dictionaries, and tuples. Lists, arguably Python's most useful data type, are similar to C's arrays. You declare a list with brackets (`[]`), and once declared, you can refer to members of a list by their index, starting with 0. For instance, type the following into IDLE:

```
spam = ["eggs", "ham", "bacon", "beans"]
```

Then type the following:

```
spam[2]
```

and you should be rewarded with this:

```
'bacon'
```

List members can be almost anything, including other lists; this is how you construct two- and three-dimensional arrays in Python. Lists are mutable, which means you can change them in place by assigning a new variable to an index (`spam[2] = "seven"`, for example). This differentiates them from strings, which cannot be changed in place, though you *can* refer to members of a string by index.

Dictionaries are similar to lists, but they have a `key:value` relationship. You can declare a dictionary by using curly braces, and then refer to its members by key. To illustrate:

```
shrubbery = {"spam":"eggs", "knight":"ni", "black":"knight"}

shrubbery["knight"]
```

returns:

```
'ni'
```

These are the main components of Python, which you'll be working with as you program.

Python Scripts

Writing Python code in IDLE is all well and good, and is a good way to practice with the language, but IDLE's main drawback is that it doesn't easily let you save your code. When you close it, it's all over.

For that reason, you'll be using a text editor to write all of your programs. There are several camps among Linux programmers regarding the "proper" text editor to use; the two most popular are Vim and emacs. If you're familiar with one of those, great. Vim is preinstalled on the Pi, and you can install emacs with a simple `sudo apt-get install emacs` in your terminal. If you're not sure, however, or don't even know what those are, fear not: the Pi also comes preinstalled with nano, a full-featured editor that's intuitive and easy to use. I use emacs, but in this book I'll refer to my code in nano for those of you using it.

If you are working on your Pi's desktop environment (either directly or via a VNC connection), you can also use the Pi's built-in Leafpad editor (Figure B-3).

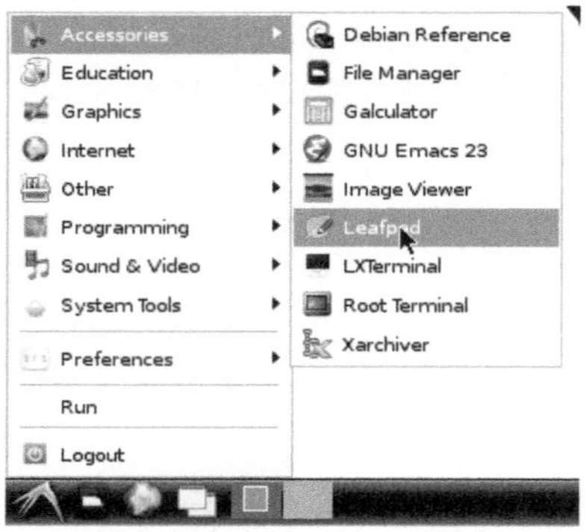

Figure B-3. *Getting to the Leafpad editor*

Unfortunately, although Leafpad will work fine for writing scripts, it can't be used from the command line. Because much of your work is done remotely after the Pi is safely ensconced in the rover, you'll have to get used to another editor.

To write a script, start your chosen text editor. From the command line, you can type **`nano test.py`**.

When you use a command-line editor like nano or emacs, your development environment will have syntax clues; that is, important Python words like `import` and `print` and `def` will be color-coded, which can be helpful when writing unfamiliar code.

Write a short script, such as this one that will print out all even integers between 1 and 100:

```
for x in range(100):
    if x % 2 == 0:
        print x
```

Now, save it as **`even.py`** and close the script. Back in your terminal, make sure you're in the same directory as the script you just wrote and type **`python even.py`**. You should be greeted by a long line of even numbers from 0 to 98.

You can make a Python script executable. In other words, clicking the file in your file manager will execute it, rather than opening it in a text editor. To do that, browse to the file's location in a terminal. From there, to change *test.py* to an executable, for instance, type this:

```
chmod 755 test.py
```

From then on, if you double-click *test.py*, you will be greeted with a dialog box like that in Figure B-4. Depending on the file's output, you can choose to run it or run it inside a terminal. This can sometimes be a time-saver if you don't want to open a terminal to enter `python test.py`.

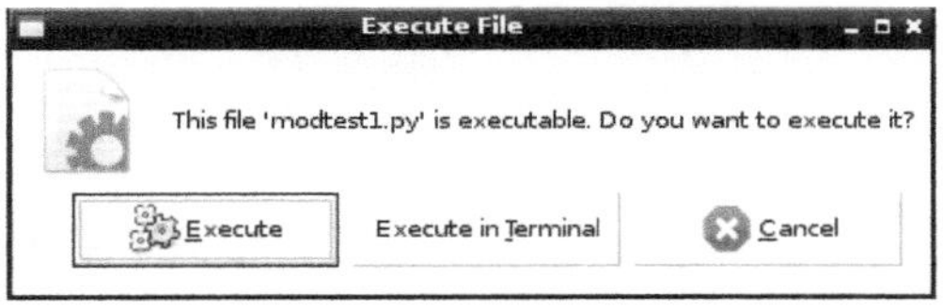

Figure B-4. *Execute File dialog box*

That is a breakdown of how to write and execute the Python scripts you'll need to program your rover. We'll come across other concepts during the build, such as functions, but I'll explain those as they arise. Hopefully, this will give you enough of an introduction to the language to let you dive in to the build!

Index

Symbols

A

B

C

About the Author

Wolfram Donat is a graduate of the University of Alaska Anchorage, with a B.S. degree in computer engineering. Along with an interest in robotics, computer vision, and embedded systems, his general technological interests and Internet expertise serve to make him an extremely eclectic programmer. He specializes in C and C++, with additional skills in Java, Python, and C#/.NET. He is the author of several books and has received funding from NASA for his work on autonomous submersibles.

Colophon

The cover and body font is Myriad Pro, the heading font is Benton Sans, the sidebar heading font is Camo Sans, and the code font is Ubunto Mono.

[redacted]

[redacted]

[redacted]

[redacted]

CPSIA information can be obtained at ww
Printed in the USA
BVOW11s1932260115

384674BV00004B/1/P

9 781457 186035